PHILOSOPHY OF PHYSICS

Other interview books from Automatic Press ♦ $\frac{\vee}{\mid}$P

Formal Philosophy
edited by Vincent F. Hendricks & John Symons
November 2005

Masses of Formal Philosophy
edited by Vincent F. Hendricks & John Symons
October 2006

Political Questions: 5 Questions for Political Philosophers
edited by Morten Ebbe Juul Nielsen
December 2006

Philosophy of Technology: 5 Questions
edited by Jan-Kyrre Berg Olsen & Evan Selinger
February 2007

Game Theory: 5 Questions
edited by Vincent F. Hendricks & Pelle Guldborg Hansen
April 2007

Philosophy of Mathematics: 5 Questions
edited by Vincent F. Hendricks & Hannes Leitgeb
January 2008

Philosophy of Computing and Information: 5 Questions
edited by Luciano Floridi
Sepetmber 2008

Epistemology: 5 Questions
edited by Vincent F. Hendricks & Duncan Pritchard
September 2008

Mind and Consciousness: 5 Questions
edited by Patrick Grim
January 2009

Evolutionary Theory: 5 Questions
edited by Gry Oftedal et al.
November 2009

Epistemic Logic: 5 Questions
edited by Vincent F. Hendricks and Olivier Roy
August 2010

See all published and forthcoming books in the 5 Questions series at
www.vince-inc.com/automatic.html

PHILOSOPHY OF PHYSICS
5+1 QUESTIONS

edited by

Juan Ferret

John Symons

Automatic Press ♦ $\frac{\vee}{\mathsf{I}}$P

Automatic Press ♦ $\frac{V}{I}$P

Information on this title: www.vince-inc.com/automatic.html

ⓒ Automatic Press / VIP 2010

This publication is in copyright. Subject to statuary exception
and to the provisions of relevant collective licensing agreements,
no reproduction of any part may take place without
the written permission of the publisher.

First published 2010

Printed in the United States of America
and the United Kingdom

ISBN-10   87-92130-32-1   paperback
ISBN-13   978-87-92130-32-7   paperback

The publisher has no responsibilities for
the persistence or accuracy of URLs for external or
third party Internet Web sites referred to in this publication
and does not guarantee that any content on such
Web sites is, or will remain, accurate or appropriate.

Typeset in LaTeX2$_\varepsilon$
Cover design by Vincent F. Hendricks

Contents

Preface

This volume is a compilation of reflections from some of today's leading physicists and philosophers of physics on the status and relevance of foundational inquiry in their field. The contributors provided responses to six relatively open questions concerning the relationship between philosophy and physics and provided their assessments of the scientific community's progress with respect to fundamental problems in physics itself.

Rather than compiling an anthology of articles, our plan was to follow the example of previous interview books from Automatic Press / VIP: We hoped to gather a collection of conversational and informal responses from luminaries in the field with relatively little editorial meddling from us. We intended our questions to elicit frank and general assessments of the foundations of physics, its scope, prospects, and future direction.

One of our goals was to convey the significance and liveliness of philosophical engagement with physics and to give readers a sense for some of the personalities involved in this enterprise. We believe that this more conversational format allows the book to serve as an accessible resource for students finding their bearings in these debates. For more seasoned scholars, the format may provide interesting new insights and at least a few genuine surprises. On more than one occasion in this volume, breaking with the standard format of academic publication has encouraged contributors to respond in more general or speculative ways than one might ordinarily expect.

⧫

A short note of explanation on the logistics of this project is in order. We posed six questions to our potential contributors and encouraged them to respond in a manner of their choosing. These are the questions we asked our participants:

1. What is the relationship between philosophy and physics? What should the relationship be?

2. How did philosophers contribute or fail to contribute to the development of physics in the 20th century?

3. What aspect of current work in physics can benefit most from collaboration with philosophers?

4. What area in contemporary philosophy of physics is most fertile?

5. In your opinion, which area of physics holds the most exciting promise in the coming decades?

6. How were you initially drawn to the field and what are some examples of your work that have influenced the discipline?

Some responded to each question in turn, giving more emphasis to some over others. Some chose to skip questions altogether and instead concentrated on what they see as the most significant issues. Notably, several of our contributors did not comment on the relationship between philosophy and physics, focusing instead on physics and its future.

The sense from most of these contributors is that recent collaborations and debates between physicists and philosophers of physics bode well for fruitful interaction in the future. Some contributors stress the need for more effective collaboration among specific types of specialists. Roland Omnès suggests that "The superposition principle is basically a mathematical statement and, along a complementary direction, I believe that a triangular collaboration between philosophers, physicists and mathematicians could be very fruitful." Richard Healey offers the following recipe for success: "Any individual seeking to contribute to our collective understanding has his or her own talents. Fortunately, the community can benefit from the contributions of people with different mixtures of abilities. This requires only that all contributors abandon arrogant professionalism and seek to combine a common spirit of sociable humility with intense and rigorous critical exchange."

Tim Maudlin reminds us that "... the most prominent physicists have generally been philosophers of physics: Newton, Einstein, Bohr, Schrödinger and Bell to name some of the most obvious. Indeed, the most profound philosophers of physics have been professional physicists rather than professional philosophers." Maudlin is undoubtedly correct to point out that the philosophy of physics is an inclusive enterprise. In spite of the influence of specialization, it is vital that discussion of foundational questions include both professional physicists and professional philosophers.

As several of the contributors note, natural philosophy as understood by Galileo, Descartes, Leibniz, and Newton had no place for a strict distinction between philosophy and physics. Insofar as foundational questions in physics are concerned, this attitude still seems to have merit. The disciplinary borders marked by departmental affiliations, journals and the like are of relatively minimal importance in these contexts. John Earman, for instance, argues that " if 'philosophy' is taken in a broader sense that is not constrained by the current disciplinary boundaries, then... when it comes to basic issues in the foundations of physics, the distinction between philosophy and physics blurs. The activity in this area is perhaps best described by a term that has fallen out of usage—'natural philosophy'." Carlo Rovelli echoes Earman when he writes that "The relation between physics and philosophy is much stronger than most physicists and most learned people generally assume today."

It has been an honor for us to work on this project and we are very grateful to our contributors for participating. We hope that readers find these responses challenging and stimulating.

Juan Ferret and John Symons
El Paso, Texas
December 2009

Acknowledgements

We are particularly grateful to the contributors for devoting time to writing such erudite, enlightening and often thought-provoking interviews and grateful to the philosophical community in general for showing interest in this project. In addition we would like to express our gratitude to Rasmus Rendsvig and Claus Festersen for preparing the manuscript with LaTeX. Finally we would like to thank our publisher Automatic Press ♦ $\frac{\vee}{|}$P, in particular senior publishing editor V.J. Menshy, for continuing to take on these 'rather unusual academic' projects.

Juan Ferret and John Symons
El Paso, Texas
December 2009

1

Frank Arntzenius

Professor

University College, Oxford, UK

---

1. What is the relationship between philosophy and physics? What should the relationship be?

2. How did philosophers contribute or fail to contribute to the development of physics in the 20th century?

Since these two questions are pretty closely related let me answer them in one go.

I don't think philosophers have contributed very much to the development of mainstream physics in the 20th century. I don't think this is a particularly bad thing, since I take it that philosophers of physics are primarily concerned with figuring out what modern physics tells us about the fundamental structure of the world, rather than that philosophers should be doing what one normally calls 'physics', which I would prefer to call 'mainstream physics'. Let me elaborate this remark a bit.

In the first place, I don't think that there is a clear distinction between physics and philosophy of physics. Rather there is a continuum of subjects that runs from applied technical physics to foundational physics, and philosophers of physics tend to be interested in the foundational end of this continuum. Philosophy of physics, at least the good part of philosophy of physics, is the attempt to develop a good theory of the more fundamental parts of physics, the parts that are very general and abstract, the parts that usually cannot be very directly tested by experiment, the parts that can significantly affect some of our most fundamental, general, and common sense ideas as to what the world is like.

In the second place, there are some areas in which philosophers of physics actually have had a little bit of influence on physics, and should have had more influence than they actually did. For example, the measurement problem in quantum mechanics for many years was dismissed by most physicists (if they contemplated such matters at all) either as non-existent, or as irrelevant, or as already solved. But there have been a bunch of philosophers of physics over the last forty years or so who have quite correctly argued that this is mistaken. An increasing number of physicists are beginning to acknowledge that this is so, and that solving the measurement problem may actually aid in developing better theories in 'mainstream physics'. While the latter may or may nor turn out to be the case, I do think it a scandal that most physicists for so long have had an intellectually bankrupt view regarding the measurement problem. It simply is a real problem and we have to solve it in order to have an even half-tenable candidate for the correct physical theory of the world. Luckily both philosophers and physicists have recently started coming up with viable solutions to the measurement problem, so we are on a better path than we have been for quite a while. A slightly more upbeat example concerns 'gauge theories'. Philosophers of physics have been concerned with the issue as to exactly what gauge theories tell us about the structure of the world, and this could very well be quite relevant when it comes to the question as to how one should quantize such theories (how one should combine gauge theories with quantum mechanics). In this area I think there actually has been some fertile interaction between philosophers of physic and physicists. The same is true of quantum gravity. So I am optimistic that the separation of Church (Philosophy) and State (Physics), which was pretty prominent in the early 20th Century, is gradually on the wane, as it should be.

3. What aspect of current work in physics can benefit the most from collaboration with philosophers?

I would expect that physicists working on quantum gravity could especially benefit from interaction with philosophers of physics. I expect this because philosophers of physics have been investigating the question as to exactly what the significance is of general covariance, diffeomorophism invariance, gauge invariance and background independence, and these issues are relevant when it comes to quantizing gravity.

4. What area in contemporary philosophy of physics is most fertile?

I don't think there is any particular area in philosophy of physics that is most fertile. Let me just list a number of recent developments in philosophy of physics that I find interesting.

A long-standing problem in philosophy of physics is how to understand the arrow of time of thermodynamics. Why is it that gases spontaneously mix, and do no spontaneously 'unmix', why is it that when two bodies are brought into contact their temperatures even out rather than that one body spontaneously becomes much hotter or colder than the other, and so on... More generally: what exactly is entropy and why does it tend to increase rather than decrease as time passes? It seems to me that this problem has pretty much been solved by David Albert (in his book Time and Chance). But there remain interesting questions in the area. For instance how is entropy increase related to other temporal asymmetries, such as the asymmetry of causation, the asymmetry of memory, the asymmetries that occur in sciences other than physics (biology, economics,...). And even within fundamental physics there still remain some things that puzzle me about Albert's account. For instance, how exactly should gravitational entropy in general, and black hole entropy in particular, be understood?

Philosophy of quantum mechanics used to be dominated by two topics: non-locality and the measurement problem. As far as non-locality is concerned I don't think there is much interesting work being done on that at the moment. Indeed it seems plausible to me that there is nothing much of interest left to be said about it: quantum mechanics is, in a clear sense spelled out by J.S. Bell and others, non-local. However, this is not a problem or a puzzle, as this non-locality is quite compatible with the dictates of Relativity Theory. The measurement problem, on the other hand, is a real problem. Luckily, nowadays there are a number of clear and potentially viable solutions available, on which interesting work is being done. In particular I would like to mention Bohmian Mechanics (especially interesting work by Shelly Goldstein and his collaborators), Many Worlds accounts (a bunch of philosophers of physics in Oxford are doing interesting work on this) and so-called 'GRW' accounts of quantum mechanics (Roderick Tumulka in particular has done interesting work in this area recently).

The philosophy of quantum field theory has recently become a much more significant area of research than it used to be. There happen to be two topics here that interest me, but this really just a matter of what I have done a bit of work on rather than that I think these are the two most fertile topics in philosophy of quantum field theory. In the first place there is a problem that is particular to quantum field theory, or rather, to any quantum theory of a system with infinitely many degrees of freedom. This problem is the so-called 'problem of unitary inequivalence'. When one has a system of finitely many degrees of freedom, e.g. a system of finitely many particles, there exists a clear and unique way of quantizing such a system, i.e. there is a method for taking a classical theory of a system of finitely many degrees of freedom and turning this into a unique corresponding quantum theory. But this method does not give a unique result when one has a system of infinitely many degrees of freedom. Consequently it is not exactly clear how one should quantize theories of systems that have infinitely many degrees of freedom. Laura Ruetsche, in particular, has done interesting work in this area.

A second topic in philosophy of quantum field theory concerns the discrete symmetries, namely charge conjugation (switching of the sign of charge), parity (spatial mirroring) and time reversal (temporal mirroring). There is a famous theorem in quantum field theory, the so-called 'CPT-theorem', which states that any local relativistic quantum field theory, i.e. one that is local and invariant under the (proper) Lorentz transformations, has to be invariant under the combination these three discrete transformations. Prima facie extant proofs of CPT-theorems do not yield much intuitive insight as to why the theorem holds. Moreover, it seems very odd that a geometric invariance (Lorentz invariance) should imply invariance under the combination of two geometric transformations (parity and time reversal) and one non-geometric transformation (charge conjugation). An obvious idea here is that charge conjugation can, after all, be understood as a geometric transformation. Hilary Greaves is doing interesting work in this area.

An area of recent interest to philosophers of physics is gauge theories. Gordon Belot jump-started this topic by distinguishing three rival ways of understanding (classical) electromagnetism. The first one is the traditional view which takes the electromagnetic field to be the fundamental quantity, the second view takes the 4-potential (the connection) as the fundamental quantity, the third one takes the loop integrals of the connection (the holonomies) as the fundamental quantities. Belot also defended the third view, and generalized it to arbitrary gauge theories. This has led to a very interesting and ongoing discussion in philosophy of physics, which also connects to the issue as to how one should quantize gauge theories in general and general relativity in particular.

Another fairly recent development in philosophy of physics that I find interesting is the revival or relationism. Relationism is the view that there is no such thing as space-time, that all that exists is material objects (which may include fields) which stand in spatio-temporal relations. Leibniz and Mach were prominent defenders of relationism, but until fairly recently the general opinion among philosophers of physics was that relationism was dead, and that some variety of substantivalism (the view that space-time exists as an object) had to be correct (in part because it was deemed hard to understand General Relativity as anything other than a theory of the properties of the thing we call space-time.) However, recent work, especially that of Julian Barbour and Gordon Belot has revived relationism as a viable option, and, as in the previous case, this issue may very well turn out to be relevant when it comes to quantizing gravity.

5. In your opinion, which area of physics holds the most exciting promise in the coming decades?

Well, I don't really know that there is much of an objective sense to be made of the notion of an 'exciting promise'. But I myself am most excited about future developments in quantum gravity, since it seems pretty likely to me that it will lead to a radically new picture of the fundamental structure of the world. I myself am attracted to the quantum loop gravity approach to combining quantum mechanics with general relativity. I am particularly excited by the idea that the fundamental structure of the world is not a continuous space-time, but rather a discrete 'spinfoam', or rather, a superposition of such discrete structures.

6. How were you initially drawn to the field and what are some examples of your work that have influenced the discipline?

From the age of ten or so I was interested in physics, not so much in applications of physics or in experiments or detailed calculations, but in what physics tells us about the fundamental structure of the world. Indeed I think that this is what initially draws most people into physics; it is what popularizations of science tend to focus on, and what famous physicists typically have found, and find, to be the most interesting. I kept this interest in the foundations of physics throughout my High School years, and then did an undergraduate degree in physics at the University of Groningen (in Holland, where I am from). Unfortunately I found that as one continues studying physics one spends more and more time doing applied physics, solving detailed technical problems, mastering certain mathematical techniques, and that one is more and more discouraged from entertaining foundational questions, since these are deemed pointless by most practicing physicists. When I insisted on asking foundational questions in my physics classes I was told, in rather exasperated tones, to go to the philosophy department instead. I duly obliged, only to discover that most people in the philosophy department at my university were primarily interested in German Continental Philosophy (Kant, Heidegger, Habermas, Adorno, Marcuse and the like) and hadn't the slightest interest in the foundations of physics. I then started reading some logical positivists (Carnap, Reichenbach,. . .) by myself and some later philosophers of science (Popper, Lakatos, . . .), and decided that I could better pursue my interests at an English University. So I went to do a Master's degree in the department of Philosophy of Science at the London School of Economics. Though better than Groningen for my purposes, it was still not ideal, as Popper had left the department, Lakatos had died, and there was no physics department at the London School of Economics. For various reasons I subsequently left academia and worked for an environmental group (Friends of the Earth) for a couple of years. But my interest in foundations of physics remained, so I went back to the London School of Economics to do a Ph.D. in philosophy of physics. I pretty much worked on my own for my three years there, though I did make some use of foundationally minded philosophers and physicists that were then around in London, Cambridge, and Oxford (in particular Michael Redhead). After that I went to the department of History and Philosophy of Science in Pittsburgh for a year as a postdoc, where I finally found people who had the same interests as I had, who were also very good and very helpful, namely John Norton and John Earman. Since then I have worked mostly in philosophy of physics, though also a bit in decision theory and metaphysics. I don't think any of my work has had any influence on the discipline, but I might have had some indirect influence by talking to people such as David Albert, Tim Maudlin and Hilary Greaves, who have had, and will have, a significant influence on the field.

2

Guido Bacciagaluppi

Centre for Time, Department of Philosophy, University of Sydney, NSW 2006, Australia and

IHPST (CNRS, ENS, Paris 1), Paris, France

1. What is the relationship between philosophy and physics? What should the relationship be?

The two disciplines have a fascinating intersection, the foundations or philosophy of physics, dealing with the conceptual aspects of physical theories, and where physicists and philosophers can and do work together closely. As a separate subject it is rather young, perhaps only 50 years old, but in recent years it has recorded some quite phenomenal progress. At least within physics, there is a growing recognition that it is an important aspect of physical enquiry (as it used to be when relativity or quantum mechanics were originally developed), but I would like to see foundations of physics become an integral part of the undergraduate physics curriculum. I also think there could be more interaction between physics (or the philosophy of physics) and more general philosophy. Insofar as physics can provide case studies that are useful for more general philosophical discussions, the lessons that can be drawn are far from straightforward, and would need more interaction to be teased out properly (say, whether and how quantum mechanics supports the notion of indeterminism). In the opposite direction, philosophy has already provided very useful tools for the study of conceptual foundations in physics (for instance, classic work on subjective probabilities and on objective chances), and could do more. In a different context, I also believe that philosophers of physics can contribute significantly to the history of physics, or at least to the history of the foundational debates within physics itself. Knowing the contemporary setting helps immensely in interpreting the historical debates, and the field is full of very rich pickings.

2. How did philosophers contribute or fail to contribute to the development of physics in the 20th century?

Some current open questions in physics, such as what makes quantum computers apparently so powerful, or how quantum theory and gravitation should be related, are directly connected to foundational issues, and the foundations of physics (whether by the hands of physicists or philosophers) could contribute interestingly to some future developments. However, for most of the 20th century physics and philosophy have led rather separate existences, and perhaps in that sense philosophers have indeed failed to contribute to various developments in physics. On the other hand, one should bear in mind that much of the philosopher's work consists in analysing the concepts used in given pre-existing physical theories (what is Newton's absolute space? what is an inertial frame in relativistic mechanics? what is probability in a many-worlds interpretation of standard quantum mechanics?), and not in the first place in helping to create new ones.

3.   What aspect of current work in physics can benefit the most from collaboration with philosophers?

I have already hinted at what I think are probably the two best candidates for this. One is research into quantum gravity, that is, the attempts to develop a quantum theory of gravitation, which classically is described by Einstein's general theory of relativity. This field raises very fundamental questions about the meaning of space and time and how they relate to quantum mechanical concepts. It is not even entirely clear which of these two sets of concepts, if any, should be taken to be more fundamental. New ideas may well be needed, and some physicists working in this field actively seek the exchange with philosophers.

The other main field in which philosophy can contribute to current work in physics is that of quantum information. This is a new field of research in physics, which has quickly acquired notoriety for its possible practical applications to computation, cryptography and so forth. But it has also incorporated many of the traditional questions in the philosophy of physics, such as the study of entanglement, non-locality and hidden variables. One aspect of quantum information that is particularly interesting for philosophers, and where philosophers have already made interesting contributions, is the study of the difference between classical and quantum theories, or the question of how quantum theories can be characterised (say, in information-theoretic terms) within some wider framework. Jeff Bub, Rob Clifton and Hans Halvorson, for instance, have proved an intriguing theorem of this kind, characterising quantum theories within the so-called C*-algebraic framework in terms of simple principles with a rather clear information-theoretic meaning ('no signalling', 'no cloning' and 'no bit commitment').

4. What area in contemporary philosophy of physics is most fertile?

The greatest progress has surely been made in the philosophy of quantum mechanics, and the field continues to produce remarkable results (although I am slightly biased of course, since this is my own field). Quantum mechanics presents us with many puzzles and difficulties of interpretation. One, if not the main, crucial problem is that, because the equation of motion of quantum mechanics (the Schrödinger equation) is linear, we cannot maintain that there is a one-to-one correspondence between the quantum states and the world as we see it. Indeed, the Schrödinger equation maps linearly superposed states to the corresponding linear superpositions of the time-evolved states. Therefore, if for instance we couple a microscopic system in such a superposition to a macroscopic system, say a measurement device, the superposition will propagate to the macroscopic level. But the world we see appears to correspond to only one component of the final superposition: there is nothing in our experience corresponding to the superposition of different readings of a macroscopic measuring device or (to use Schrödinger's famous example) a live or a dead cat. This - and a cluster of related problems - is the quantum measurement problem.

Within the field of foundations, there have traditionally been three principal approaches to quantum mechanics: pilot-wave theories, collapse theories and many-worlds interpretations, all of which provide modifications or interpretations of quantum mechanics that arguably do not suffer from the measurement problem. All of these have been radically transformed in recent years. In addition, we are now witnessing the beginnings of possible new approaches that are more epistemic, or 'information-based'. Let me say a few things on these four kinds of approaches in turn.

Pilot-wave theories (such as the theory developed by Louis de Broglie in the years leading up to 1927, and again by David Bohm

in 1952) are based on the idea that the quantum state determines
the velocities of point particles (or the dynamics of some other
suitable configuration variables). As such, they are the prime example of so-called hidden variables theories, in which the description of a system includes something - here particle positions - beyond the quantum state. (This something determines in particular
the measurement results.) The theory was revived in the 1980s by
students of Bohm, such as Peter Holland and Chris Dewdney, and
it has undergone major advances in the last 15 years, in particular through the work of Antony Valentini on the one hand, and
of the group of researchers around Detlev Dürr, Shelly Goldstein
and Nino Zanghì on the other. Some of these advances include:
the suggestion and ongoing elaboration of an analogy with statistical mechanics, the study of non-equilibrium solutions (which
provide the possibility of experimental disagreement with quantum mechanics), and various new models for field theory, of which
the most exciting are the very recent ones by Samuel Colin and
others, and by Ward Struywe and Hans Westman. From these it
now appears that pilot-wave theories, although they are arguably
non-relativistic at the level of the hidden variables, can indeed
reproduce all the predictions not only of non-relativistic quantum mechanics, but of the standard model of elementary particle
physics as well.

Collapse theories are the most recent of the three more traditional approaches to quantum mechanics, and they modify the
theory by proposing modifications to the Schrödinger equation,
usually stochastic ones. That is, detailed mechanisms are worked
out that in typical measurement situations will force the quantum state to evolve to any single of the macroscopically observed
alternative final states, with probabilities matching the usual statistical predictions of quantum mechanics. Collapse theories have
been recognised as a viable approach to the foundations to quantum mechanics since a famous paper by GianCarlo Ghirardi, Al-
berto Rimini and Tullio Weber from 1986, but the work by Philip
Pearle both before and after was also crucial for the development
of the approach. The main problem faced by collapse theories
(as in the case of pilot-wave theories) was originally the apparent
incompatibility of the collapse mechanism with relativity. Indeed,
collapse appears to be a process occurring across the whole of
space. So, for instance, coupling a multi-particle state to a measurement device situated at a particular location will induce not
only a collapse of the state of the particle that interacts with the

device directly, but a collapse of the total state, including particles that will be detected at arbitrary distances from the first one. But it now appears that the original idea of how a quantum state fits into a space-time context was mistaken. Thanks to recent philosophical analysis by Wayne Myrvold, it is now becoming accepted that the general notion of collapse is perfectly compatible with relativity. The problem is to define a specific collapse mechanism that will be relativistically invariant, and some first models along these lines have also recently been constructed, in particular by Rimini and co-workers.

Many-worlds interpretations originate from the 1957 seminal paper by Hugh Everett. (A great deal of interest in Everett and the genesis of his ideas has been sparked off by the 50th anniversary of this publication.) The central intuition behind these interpretations, at least as understood today, is that the quantum state of the universe contains dynamically independent components that all describe different 'worlds', each containing for instance a different result of a quantum mechanical measurement, a dead or live Schrödinger cat, and so forth. The motivation for adopting such an interpretive strategy is that in a many-worlds interpretation, properly construed, there would be no objects apart from the quantum state itself, and no deviations from the Schrödinger equation, with the further advantage of preserving compatibility with relativity (once one applies the interpretation to a relativistic quantum theory). How all of this is supposed to work remained obscure for a long time, although the idea of many-worlds gained much support from quantum cosmologists after being championed by Bryce DeWitt in the 1970s. It is only very recently that a version of the many-worlds interpretation has emerged that is both well worked-out and does what it is supposed to do. The turning point for many-worlds interpretations was the realisation that the dynamically independent structures that constitute worlds arise through the mechanism of decoherence, that is the lack of interference over time between components in certain decompositions of the quantum state at a certain level of description. This realisation emerged mainly through the work of Simon Saunders, who from the early 1990s onwards developed a detailed account of how to make sense of Everett worlds - partly based on the analogy with the so-called block universe picture of time - and of how worlds (and observers) 'split'. (To be fair, Everettian ideas had been associated with decoherence already by some of the main workers in the theory of decoherence, in particular Hans-Dieter Zeh and also Wojciech Zurek.) The last years have further seen the development of a proposal for understanding what probability means in an Everett universe, namely in terms of decision theory as applied to splitting agents. This was principally through the work of David Deutsch and of David Wallace (and of Hilary Greaves). The result is a particularly good illustration of the close interaction one can establish between philosophy of physics and other branches of philosophy. It is the perfect embodiment of David Lewis's Principal Principle, not only in the sense that the Principle provides the strategy for introducing probabilities in an Everett universe, but also in the sense that the proposed solution provides the arguably cleanest example of how the Principle can be used to identify objective chances out there in the world.

The most recent among the foundationally interesting approaches to quantum mechanics are related to the development of quantum information. The crucial aspect of these approaches (or of these strategies for developing new approaches), I think, is the idea that the quantum state, instead of being an ontic notion, should be analysed in epistemic terms. For instance, in the approach that is being developed by Chris Fuchs and others, the central idea is that, instead of being a compendium of objective probabilities, the quantum state is a compendium of subjective probabilities, much in the sense of De Finetti. The whole apparatus of subjective Bayesianism (with suitable generalisations) is brought to bear on quantum mechanics, and this strategy may indeed turn out to be very fruitful. With hindsight, this is an obvious thing to try, but it is a new development, which provides a novel entry into the question of what differentiates between classical and quantum theories. This approach also provides a modern version of ideas that are often associated with various historical exponents of the Copenhagen interpretation of quantum mechanics. In particular, if quantum states are interpreted as subjective probabilities, this makes it unremarkable that one should require the subject to play a special role in the theory. (The approach faces some of the same challenges, however.) Another example of work on epistemic approaches is the 'toy theory' by Rob Spekkens, which is a stunning piece of work, showing how dozens of features known from quantum mechanics or quantum information can be reproduced in a theory that is entirely classical at the level of the ontic states, but that postulates limitations on the epistemic states. Again this is an important step towards recognising what makes quantum mechanics really quantum.

5. In your opinion, which area of physics holds the most exciting promise in the coming decades?

Research into quantum gravity certainly appears to call for potentially radical changes in present physics. Traditional techniques for quantising classical theories, as have been used for instance in constructing quantum electrodynamics, have helped only up to a point in trying to construct a quantum theory of gravity. In fact, general relativity (at least in its usual formulation) is very far from being a standard field theory. Field theories assume a background spacetime, while general relativity treats simultaneously both spacetime and the matter fields on it as dynamical objects. It is not even clear whether a microscopic theory of gravity is necessary (which is the usual assumption), or whether gravity might emerge only on a larger scale as an effective theory. These may be signs that some big conceptual changes are needed.

6. How were you initially drawn to the field and what are some examples of your work that have influenced the discipline?

I was an undergraduate in mathematics and physics at ETH (the Swiss Federal School of Technology) in Zürich. One of the many good features of that university was that, every semester, each student had to enroll also in a humanities course (presumably a left-over from an enlightened 19th-century attitude towards education). I was in my first semester of study, and one of my best friends tipped me off about the lectures and seminars of a guest professor from Boston, one Abner Shimony. As a matter of fact, Abner was holding a lecture course on naturalised epistemology, as well as a seminar on foundations of quantum mechanics (organised by Hans Primas). I think I missed the first session, but from then onwards my friend (Thomas Breuer, together with whom I also later trained in philosophy at Cambridge) and I attended enthusiastically all the lectures and seminars, and soaked in everything that Abner had to say. We also had the opportunity to attend a number of sessions with invited speakers: a young Nicolas Gisin, a beardless Anton Zeilinger, and the great John Bell himself (on his new 'beable' theories, which in fact became the basis for part of my Ph.D. thesis). Not that we (or I at least) understood very much at the time. It was only weeks later that, for instance, the notion of the trace of a matrix came up in our linear algebra lectures (taught by Ernst Specker)! But Abner (with whom I have kept in touch ever since) was splendid, and the damage was done. Later on, when I was wondering whether to become a professional mathematician or do something else, it was again Thomas who tipped me off about Michael Redhead's group in Cambridge. Abner sent off a 'strong letter of recommendation', and by the time Thomas and I had visited Cambridge and had been received by Michael in what was then his office in the HPS department, with Jeremy Butterfield and Rob Clifton as heraldic supporters, and surrounded by their already large research group, the decision was made ('I think we'll take them... '). I was to spend five happy years as a graduate student in Cambridge (which really meant being treated as a colleague), under the guidance of Michael first, then Jeremy. That long-lost research group has always remained the pinnacle of my experience of a research setting, and I am deeply indebted to both Michael and Jeremy (and to Abner before them) in more ways than I can say.

My early work, during my Ph.D., was mainly on modal interpretations of quantum mechanics (specifically what was known as the Kochen-Healey-Dieks version), and I certainly had an impact on the community of researchers that was working on that topic. With Meir Hemmo, I provided what at the time seemed a solution to the 'problem of imperfect measurements' raised by Albert and Loewer, using tools from the theory of decoherence. I proved a Kochen-Specker theorem for the modal interpretation, which constrained the further development of the theory. Together with Michael Dickson, I developed a dynamics for the 'hidden variables' of the interpretation, by generalising Bell's stochastic 'beable' dynamics to the time-dependent case. And I showed (prompted by Matthew Donald's vigorous criticism both in discussion and in print) that if decoherence is fully taken into account, problems similar to the Albert-Loewer one come back with a vengeance. As a consequence, the modal interpretation (at least in the version I was working on) fails to meet the requirements of empirical adequacy. That may in fact be a reason why interest in these interpretations has waned (certainly my own interest in them). On a more positive side, I believe that my work on decoherence has had a fair impact on the way philosophers of physics see the role this notion plays in the foundations of quantum mechanics, specifically in 'no-collapse' approaches to the theory. I also hope that my work as subject editor in quantum mechanics for the Stanford Encyclopedia of Philosophy may be helping to spread the understanding of the field more widely. And I believe very much in furthering the field through interaction and collaboration with young researchers and graduate students, as my teachers did with me.

3

Mario Bunge

Department of Philosophy
McGill University, Canada

The Troubled Relationship Between Physics and Philosophy

1. What is the relationship between philosophy and physics? What should the relationship be?

Science and philosophy were rather close from the Scientific Revolution to the Enlightenment. So close, in fact, that some of the great brains of that period– particularly Galileo, Descartes, Huygens, Boyle, Euler, and d'Alembert–were philosopher-scientists. Indeed, in addition to making momentous special discoveries and inventions, they built the modern mechanist ontology–the earliest science-based materialism. And, with the exception of d'Alembert, an early positivist, they also defended the realist epistemology against the phenomenalism and conventionalism then favored by the Catholic Church.

But even during that time there were some eminent philosophers, such as Locke, Berkeley, Hume, Vico, Kant, and Rousseau, who ignored or even opposed the science of their time. For example, Locke, Berkeley, Hume, and Kant declared that only phenomena (appearances) could be known; and Vico and Rousseau, early Romantics, were just as hostile to science as Berkeley. Except for Berkeley, who subjected the infinitesimal calculus to intelligent criticisms, none of those philosophers had the mathematics required to understand and appreciate Newtonian mechanics. Yet some of them, in particular Hume, dared criticize it. And Kant did not deign to reply to the letters that Johann Heinrich Lambert, the brilliant polymath, wrote him criticizing his repulsive force and his subjectivist views of space and time. By contrast, Leonhard Euler, Newton's heir, did not waste his time: he called the subjectivists 'clowns', and noted that the watchdog who barks at a stranger is no subjectivist.

The German romantics managed to worsen the relations between science and philosophy. They rejected science and invented their own lunatic Naturphilosophie. In particular, Hegel dismissed all the most advanced scientific theories of his time, including Berzelius' prescient explanation of the formation of ionic compounds. Perhaps the only scientist to take notice of such wild fantasies was Ampère, the founder of electrodynamics and author of a treatise on the philosophy of science. In this work he criticized Schelling's fantasies, and drew the critical difference between scientific laws and laws of nature–the objective patterns that the former attempt to capture. If philosophers had taken notice of Ampère's philosophical foray, they would not have written about the inaccuracy of the laws of nature, even less about their mendacity. Acquaintance with the theory of errors, fathered by Gauss nearly two centuries ago, might have had the same effect. But philosophers have characteristically ignored this theory.

However, the hostility between the two camps is far more comprehensible than the case of the many eminent scientists who unwittingly adopted philosophical views that undermine the scientific enterprise. This was the case of such distinguished and influential physicists as Ernst Mach, William Rankine, Gustav Kirchhoff, Wilhelm Ostwald, Pierre Duhem, and others. They all rejected atomistics in the name of positivism, an heir of the phenomenalism of Ptolemy, Hume, Kant, and Comte. (Duhem had also been strongly influenced by Thomism, to the point that he proposed going back to Aristotle's qualitative physics.) Only Ludwig Boltzmann, the founder of statistical mechanics, resisted the ruling positivist philosophy and defended not only atomistics but also materialism. And ironically Lenin, a political activist and amateur philosopher with no scientific education, attacked the popular idealist reading of physics in an interesting book read only by Marxists: He defended the right stand (realism) for the wrong reason (because Engels dixit).

The resistance of those classical physicists to the atomic theory might be excused because the first solid evidence for the existence of atoms was produced only in 1905, in the light of Einstein's paper on Brownian motion. But phenomenalism is puzzling in the founders of modern atomic physics, particularly Bohr, Heisenberg, Born, Pauli, and Dirac, for they attempted to interpret quantum mechanics, which handles unobservable things–such as photons and quarks, and unobservable properties, such as parity and atomic energy levels–in terms of observables that are not referred to by the theory. This was the philosophical gist of what became known as the Copenhagen interpretation of quantum mechanics.

Between the two wars Bohr's institute, generously financed by the state brewery, became the Mecca of quantum physicists. Among others my teacher, Guido Beck (1903-88), made his yearly pilgrimage to Copenhagen, to try and decipher Bohr's cryptic comments on his favorite formulas, "$E = h\nu$" for photons, and "$\lambda = h/p$" for particles. But the Copenhagen Oracle had earned his reputation for obscurity. Half a century later, my teacher confided in me that, whereas he understood the two relativities, he had never really understood the quantum theory. My complaint was that we, the foot soldiers, were expected to obey a slight variation on Mussolini's famous commandment: "Believe, obey, calculate." The philosophers could not help: if interested at all in the new theory, they were too busy to learn it, and happy to embrace uncritically the ready-made Berkeley-Hume-Kant-Comte-Mach philosophy inherent in the Copenhagen interpretation.

True, Planck and Einstein, and occasionally Schrödinger and Louis de Broglie as well, criticized the Copenhagen obscurities and inconsistencies, and defended realism. But they did not offer a detailed realist alternative. (Pauli, who was said to wreck all the experiments he came near to–whence the popular phrase "Pauli effect"–pointed out that the critics had no theory of measurement.) Besides, they often conflated realism with classicism, in particular causal determinism.

So, the critics were ignored and sometimes even ridiculed as dinosaurs: Berkeley's dead philosophy proved stronger than the realism that should accompany the exploration of the physical thing in itself, endowed only with primary qualities, as Galileo and Descartes had taught. Needless to say, the Vienna Circle (né in 1928 as Ernst Mach Verein) and its affiliates, in particular Philip Frank and Hans Reichenbach, was solidly behind the Copenhagen philosophy.

The philosophers' reaction to the special theory of relativity (1905) had been far simpler. Henri Bergson and other intuitionists rejected it out of hand because it hurt their pre-scientific intuitions of space and time–which, like Kant, they took to be anchored in the human mind. Others accepted the theory because they thought it was observer-centered, that is, subjectivist or relativist in the philosophical sense. (The famous sociologist of

science Bruno Latour was one of the last defenders of this confusion between the two senses of 'relativity'.)

Einstein and Planck protested repeatedly against the confusion
between reference frame and observer, emphasized that relativity to a reference frame has nothing to do with subjectivity, and
argued that the theory uncovers new invariants instead of con-
firming the popular view that everything is relative. But Einstein
himself unwittingly contributed to the said misinterpretation in
his popular writings: he should have used photocells instead of
human observers. The only philosopher who understood correctly
the theory was Emile Meyerson, one of the two philosophers who
exchanged letters Einstein–the other being Moritz Schlick, who
had been a realist before founding the Vienna Circle. (Michele
Besso, Einstein's close friend and colleague at the Patents Office,
and who dabbled in philosophy, tried unsuccessfully to convert
him to positivism.)

In short, the relation between physics and philosophy after the
Scientific Revolution was largely a comedy of errors. Philosophers either ignored scientific advances or, which is far worse,
succeeded in persuading some of the most innovative scientists
that their scientific work had confirmed the phenomenalism of
Berkeley, Hume, Kant, Comte, Mill, and Mach, which ignored
or even outlawed the boldest research projects–particularly field
physics, atomistics, and astrophysics–for wishing to unveil the re-
alities lurking beneath appearances.

Such incongruities might have been prevented if philosophers
and scientists had taken each other seriously: if they had bothered
to examine critically what they were accepting or rejecting out
of hand. For example, if instead of demeaning all philosophy
he had sought the company of pro-science philosophers, Richard
Feynman might not have stated confidently that the electron is
just a useful theory (his emphasis): he would have learned that
theories have no physical properties, just as electrons have no
conceptual ones. And he would also have learned that, for all his
contempt of philosophy, he had unwittingly adopted one, namely
instrumentalism, which is hostile to the disinterested research he
practiced and defended.

In conclusion, on the whole the physics-philosophy relationship
has been disastrous: from indifference to the prevalence of a philosophy eager to impose strictures on scientific research. Yet co-
operation between the two is possible provided the philosophical
partner talks clearly rather than postmodern gobbledygook, and

embraces an ontology and an epistemology inspired by the scientific exploration of the world. In short, interaction among equals is the ticket.

2. How did philosophers contribute or fail to contribute to the development of physics in the 20th century?

All of the 19th century philosophers ignored the most important physical revolution of their own time: the birth and development of field physics and, with it, the start of the decline of the mechanist ontology that had prevailed since Galileo and Descartes. By contrast, the main offspring of classical electrodynamics, namely the special theory of relativity (1905), drew the attention of nearly everyone–except of course Thomists, Hegelians, phenomenologists, and existentialists. In particular Heidegger's Being and Time, published two decades after the birth of relativity, does not mention it; worse, it "defines" time as "the maturation of temporality." A few Neo-Kantian philosophers, particularly Cassirer, attempted unsuccessfully to repair the considerable damage that Einstein's theory inflicted upon Kant's view that space and time, far from being objective, are anchored in the human mind.

The influential Henri Bergson and other intuitionists rejected the theory because it hurt their intuitions of space and time. (Bergson ended up by silently withdrawing his offending book from circulation.) Other philosophers accepted the theory because they thought that it was observer-centered, that is, subjectivist or relativist in the philosophical sense. The dialectical materialists rejected relativity for the same reason: because they misunderstood it as subjectivist. Fortunately the eminent Russian physicist V. A. Fock defended the theory and succeeded in having the ban on it lifted. The Soviet philosophers believed that quantum mechanics too was idealistic, and accordingly fought it. This time the eminent Lev Landau came to the rescue. Later on quantum chemistry too came under philosophical attack. These episodes persuaded the Soviet physics community that philosophers were ignorant enemies to be shunned. At the Relativity Conference in London, in 1965, I met some physicists with the Lumumba University in Moscow, who told me that they held a philosophy of physics seminar, but did not allow any philosophers to attend it.

The reaction of philosophers to quantum physics was quite different. Most of them refrained from making any pronouncements about it, perhaps because it cannot by popularized (and distorted) by describing train passengers measuring distances and time intervals. Besides, the theory seemed to come equipped with its own philosophy, namely operationism. This offshoot of positivism and pragmatism had been "in the air" since the turn of the century, and it was expounded in 1927 by Percy W. Bridgman, the great experimentalist–who claimed, among other things, that the expression "speed of light" is meaningless, for light can only be detected when emitted or absorbed, never in between.

Werner Heisenberg, the Mozart of physics, espoused operationism and claimed that his own matrix mechanics of 1925 complied with it: that it contains only variables denoting directly observable quantities. If a critical philosopher had been at hand, he would have remarked that, since Heisenberg's p and q matrices are infinite, they cannot stand for observables. Likewise, when Louis de Broglie and Erwin Schrödinger introduced wave mechanics a couple of years later, it should have been pointed out that its key function, the famous Ψ, was anything but observable.

Actually the only quantum theory that obeys the operationist recipe is Heisenberg's S(cattering) matrix theory of 1943, which purports to describe every microphysical object as a black box with only two terminals: input and output, both measurable in collision experiments. But the S matrix was soon shown to be parasitic on standard quantum mechanics. Its father does not even mention it in his autobiography. And in 1969 Heisenberg told me that, contrary to his young colleagues, always eager to bow to the latest experimental data, he was of a Newtonian cast of mind, in that he wanted to understand facts, not just to describe them. (And of course for a physicist understanding requires conjecturing the mechanism(s) transducing inputs into outputs. The standard "model" of explanation, as deduction from law and circumstance, overlooks mechanism, or that which makes things "work".)

Unsurprisingly, most philosophers of quantum mechanics accepted uncritically the philosophical musings of Bohr, Heisenberg, Born, Pauli, and other great physicists, believing that their philosophy resulted from their scientific research. And in most cases philosophers focused on a few formulas, such as Heisenberg's inequalities, or on a single experiment, typically Stern-Gerlach's. They ignored that disclosing the meaning of even a familiar concept, such as that of mass, calls for a whole set of propositions involving the given concept. Yet none of those philosophers would judge a philosophical system, or even a person, by a single trait.

Furthermore, most philosophers consistently perpetrated the original sin of operationism: They confused meaning (in particular reference) with empirical evidence. (After all, they read, even in the Encyclopedia Britannica, that relativity is about measurements of distances and time intervals, and that the quantum theory is about diffraction observations and spin measurements.) This is how so many inconclusive papers were written about the so-called particle-wave duality and Bohr's complementarity principle.

The shift of focus, from things to the ways they are observed or measured, from reference to evidence, prevented people from realizing that the things the quantum theories describe are neither particles nor waves: instead, they are non-classical objects and, as such, deserving of a non-classical name. Four decades ago I proposed to call them quantons, but this suggestion did not prosper. I also inveighed against the abuse of analogy and, in particular, against the view–pioneered by Max Black and Mary Hesse–that analogies are the cores of theories other than theories about analogy. Thus, under certain circumstances quantons are similar to particles, whereas under different circumstances they look like waves. Ironically, Heisenberg himself had adopted a similar view in his Chicago lectures of 1930. But Bohr's view prevailed: He loved chiaroscuros, as his coworker Leo Rosenfeld reported. (When asked what was complementary to truth, Bohr is said to have answered: Depth.)

Still other philosophers did more than join the bandwagon: They contributed to two offshoots of the Copenhagen credo, namely quantum measurement theory and quantum logic. In my opinion neither of these fields has been fruitful. The reason for the failure of the former theory is that the measurement of any physical magnitude calls for special instruments, the design and operation of every one of which calls for its own special theory (as Duhem had noted long ago). For example, whereas the spectrograph does not alter the position or breadth of spectral lines, the ammeter consumes energy, and thus alters the intensity of electric currents. (Even Wolfgang Pauli, a passionate defender of the Copenhagen creed, admitted the existence of non-invasive measurements.) Only a theory of the instrument in question can suggest and justify the use of the corresponding indicators. If in doubt, ask any of the physicists or engineers at CERN which of their measurements–which nowadays costs around one billion dollars–has been designed with the help of the theory in question.

As for quantum logic, it suffers from both a severe birth defect and impotence.

The former is the confusion, which occurs already in the Birkhoff-von Neumann founding paper, between operators and propositions. And the impotence in question is obvious from the fact that quantum logic has not produced a single empirically significant theorem since its birth seventy years ago. Four decades ago I asked Werner Heisenberg whether he had ever used quantum logic. His reply was revealing: "I know nothing about quantum logic. Ask von Weizsäcker, he knows all about it." But, although Heisenberg's star pupil wrote many papers on the subject, he never obtained any physical result with the help of that logic. My verdict is that this field, just like the quantum theory of measurement, is just an academic industry.

In sum, in my opinion the standard philosophy of physics has not made any constructive contribution to physics in the 20th century. Worse, most philosophers have ignored the arguments of Galileo and Descartes about primary and secondary qualities, and have preached a return to Ptolemy's injunction, "Stick to appearances." In other words, they have been scientific and philosophical reactionaries.

3. What aspect of current work in physics can benefit the most from collaboration with philosophers?

All of physics can benefit from proscientific philosophy, just as it can be hurt by bad philosophy. But some fields can benefit most: macrophysics and mesoscopic physics, quantum physics, string theory, cosmology, and quantum chemistry. Let me explain briefly.

Philosophers can point out that it is mistaken to leave the study of macrophysical and mesophysical systems to engineers, as has been the case for the past half century, for these are usually under pressure to produce useful results. The fact is that we still do not have a good theory of liquids, and that mesophysics is still embryonic. Mesoscopic facts are particularly interesting because they occur at the micro-macro interface, and thus call for both classical and quantum physics, and raise an important if neglected ontological problem: How do macroproperties (bulk traits) emerge? For instance, which is the smallest physical system endowed with thermodynamic properties, such as temperature and entropy; and how many photons does it take to form a laser ray?

Quantum physics would benefit enormously from a closer contact with ontology–without which philosophy is invertebrate. Take, for instance, the concept of individuality, long regarded as an essential trait of all objects, in particular material things. The universality of individuality was first challenged by quantum statistics, both Fermi's and Bose-Einstein's. Indeed, both assume that elementary particles are equivalent, hence exchangeable. Indeed, although they can be numbered, they cannot be tagged. (An exchange of the position variables in the state function leaves it invariant for integral-spin particles, and only changes the sign of the function for the others.) This is why the elementary particles are commonly said to be identical, while the correct word is 'equivalent'.

The second challenge to the principle that everything can be individuated was entanglement. Its discovery maimed the individualism inherent in ancient atomism, nominalism, and methodological individualism: it showed that "once a system, always a system." In other words, it is no longer possible to study isolated individuals: they have to be placed in context. A cartoon in a recent issue of The New Yorker shows this graphically: A lady complains to her mechanic that her car starts moves smoothly in the driveway, but gets stuck in traffic. She had not heard that every thing in the world is either a system or a component of some system. And she is not alone in this ignorance: the very concept of a system does not even occur in the standard philosophy dictionaries. When suggesting earlier that physics would benefit from a closer contact with ontology, I had scientific ontology in mind, not mainstream metaphysics, which wanders lost in the wilderness of possible worlds.

String theory, which is being touted as the theory of everything, and is employing thousands of physicists and mathematicians, has been in existence for exactly four decades. Any critical philosopher of physics is bound to ask some embarrassing questions, such as the following. One of the truth tests that any new theory has to pass is what I have called external consistency: it must agree with the bulk of the fund of knowledge. But string theory fails this test, for it postulates that physical space is 10-dimensional, whereas according to all the other theories it is 3-dimensional. True, no particular dimensionality shows up when writing physical equations in Cartesian coordinates. But it becomes clear when adopting spherical coordinates: the second radial term of the laplacian contains the expression "$n - 1$", where n designates the number

of spatial dimensions. The occurrence of the 7 extra dimensions
has been "explained" by some ad hoc hypotheses, but these sound
like so many excuses. One of the latest and most extravagant is
that the universe was initially ten-dimensional, but eventually it
lost the 7 extra dimensions through an unknown mechanism–use
it or lose it?

Cosmology is another area that could benefit from philosophical questioning. This is because philosophers have always been
fascinated by questions of origin, and because all such questions
are likely to elicit speculation beyond the data at hand, and correspondingly to look for very indirect evidence. A critical philosopher is likely to ask some embarrassing cosmological questions.
For example, if the Big Bang did in fact occur, what was it that
exploded: what were the properties of pre-bang matter? And
what caused the explosion, that is, what was its mechanism? A
critical philosopher will stick to Lucretius' ex nihilo nihil principle: He will reject as theological any explanation involving the
assumption that the universe originated out of nothing. He will
not even accept the conjecture that the universe emerged from a
pre-existing quantum vacuum, for this would involve the sudden
creation of energy–a no-no fantasy. Last, but not least, a critical
philosopher will warn cosmologists not to boast over their latest
speculations. Some humility is in order given that megaphysics is
so young and immature, that until recently it was not known that
90% of the universe is occupied by "dark matter", that is, matter
that is neither luminous nor reflecting, which suggests that it is
quite different from ordinary matter.

Finally, quantum chemistry is a nearly virgin subject of the
philosophy of chemistry, a field that is gaining adepts by the day.
(I first met it in 1947, when the Argentine chemist Carlos Prélat
published one of the earliest books in this area: Epistemología
de la química.) One of the main open problems in this field is
whether the reduction of chemistry to physics has been or can be
total (mere reduction); or, as I believe, it is partial (deduction
from quantum mechanics enriched with subsidiary assumptions).
Regrettably, the philosophers of science have been somewhat re-
miss in analyzing real cases of reduction. (In the early 1950s
Philosophy of Science rejected my first paper on this problem, for
challenging the popular view that thermodynamics had been fully
reduced to mechanics.)

In short, philosophers can ask some fruitful if sometimes upset-
ting physical questions, which may induce physicists to moderate

their chronic triumphalism.

4. What area in contemporary philosophy of physics is most fertile?

I do not know which area is most fertile, but I have a strong opinion about important areas that have been unduly neglected. Let me list a few: axiomatics; a deeper analysis of quantum field theory and quantum statistical mechanics; exploration of the micro-macro interface; and an investigation of the indicators that mediate between testable theorems and the corresponding empirical data. Let us glimpse at the first task.

Axiomatization is indispensable to avoid discussions out of context, which discussions typically overlook the most important, namely the things that one is talking about. For instance, we would have been spared much nonsense if the most widely exploited formulas, such as "$E = mc2$", and "$\Delta p \cdot \Delta q \geq h/4\pi$", had been not been discussed out of context. Einstein's would have been discussed as a theorem in relativistic mechanics rather than as the universal equivalence of mass and energy. And Heisenberg's would have shown that the so-called uncertainties are actually objective indeterminacies unrelated to measurement, since they follow from very general assumptions. A second reason for urging further work in axiomatics is that it alone shows the need to enrich a mathematical formalism with semantic hypotheses (the old "correspondence rules"), such as the one that specifies the kind of thing a theory describes. Norman Campbell's advocacy for such extramathematical assumptions in 1919 was ignored, partly because physicists usually make them tacitly, usually by waving their hands.

5. In your opinion, which area of physics holds the most exciting promise in the coming decades?

I will resist the temptation to issue prophecies. Who knows where the most amazing novelties will emerge? Recall that, prior to 1912, nobody expected anything spectacular from Einstein's work in a subject that was regarded as dead–gravitation; that in 1935 nobody foresaw that Schrödinger's cat would become the most famous cat in history; that before WWII no one foresaw that quantum mechanics would help give birth to the digital computer;

or that the quantum-theoretic modeling of molecules and chemical reactions would turn out to be tremendously difficult.

6. How were you initially drawn to the field and what are some examples of your work that has influenced the discipline?

When waking up intellectually, I became fascinated by psychology, philosophy, and astrophysics– all of which I had, of course, only glimpsed. I got over my infatuation with psychology by writing a book-long criticism of psychoanalysis, just before entering university. (Fortunately the manuscript got lost.) I enrolled as a physics student because I wanted to refute the idealist philosophies of physics that I had read in the popular writings of the distinguished physicists Sir Arthur Eddington (a Kantian) and Sir James Jeans (a Platonist). I studied philosophy on the side and on my own. In 1939 I gave my first two public lectures on philosophy; unfortunately they were published.

In 1943 Guido Beck, an Austrian refugee who had been an assistant of Heisenberg's in Leipzig, took me under his wing and put me to work on nuclear forces. (My first physics papers appeared in 1944, and the latest in 2003.) The next year I founded the philosophical journal Minerva, whose main goal was to counteract the irrationalist wave that came from Germany and France. I regarded this as my contribution to the war. (Argentina, my country, remained neutral in the war, but the government was pro-German.) My teacher's reaction was to drop me: He rightly insisted that scientific research demands full-time dedication; and, like most scientists, he felt contempt for philosophy. But I went on calculating and reading physics literature, and sketched a theory of the radiation that Cherenkov had recently discovered. This effort was unsuccessful, but it earned me Beck's resumption of his supervision.

Beck proposed for my Ph.D. dissertation a problem of some philosophical interest, namely to find out whether relativistic quantum mechanics contained Bohr's semiclassical model as an approximation. This work involved tackling a number of divergent integrals. My two weeks in jail, where I was deprived of paper and pencil, helped considerably, for I had to do mental calculations, which sharpened my intuition about those devilish integrals. Back home I finished my dissertation, which was published in book form only a decade later because I had become an outcast. I was not even allowed to attend the graduation ceremony, and lost my university job.

In 1953 David Bohm, who had liked my paper on "What is chance?," invited me as his postdoctoral fellow at the Sao Paulo University, where he was weathering the McCarthy storm. My discussions with him gave me the idea of writing two books, one on causality and the other on levels of organization. Both were motivated by the attempt to understand the quantum theory. Eventually, back in Buenos Aires, I wrote both books over the course of the next five years. Harvard University Press published the former in 1959, and it was an instant hit. By contrast, no sooner did I complete my book on levels, than I discovered that, misled by ordinary-language philosophy, I had conflated a number of different concepts of level, and consequently consigned this project to oblivion. Eventually I picked the concept of level that had been popularized by physicists and biologists, and exactified it. This concept, together with that of emergence introduced in 1879 by George Lewes, was to become one a cornerstone of my ontology.

During those years of forced exile from academia, some of us, while working in the private sector, held a weekly physics seminar, where we discussed our work and some of the papers in the literature. I also led an underground philosophy seminar, only one of whose members was a philosopher: the rest were scientists, a mathematician, and a couple of physicians. No such seminars were held at the university. The 1955 coup that deposed the Perón government put an end to those delightful meetings. A year later I was appointed a professor of theoretical physics at the universities of Buenos Aires and La Plata, and in 1957 I won the philosophy of science chair at the Universidad de Buenos Aires. In 1963, fearing a new military coup, I left Argentina for good. I taught both physics and philosophy at Temple University in Philadelphia, and the next year at the University of Delaware.

While teaching quantum mechanics at Temple, and the two relativities at Delaware, I conceived of my next project: the axiomatization of all the fundamental physical theories. This project was one of the 23 open problems that David Hilbert had formulated in his famous 1900 address. I completed this project at the Institute of Physics at the University of Freiburg, under an Alexander von Humboldt research fellowship. My work on this project is contained in my Foundations of Physics (1967). A single philosopher reviewed it. Because he had no ear for mathematics, he was unable to make any specific criticism—which did not prevent him from damning the book in its entirety.

My main contribution to physical axiomatics was the explicit addition of semantic assumptions to the mathematical postulates. I couched those assumptions in realistic terms, avoiding any reference to observers and empirical operations, which I exiled to the laboratory. The typical semantic assumptions are of the forms "S is the set of all things of kind K", and "Function (or operator, or element of an algebra) M represents physical property (or else event) P".

Framing such assumptions can be tricky in the case of abstract and unfamiliar formulas. For example, the symbol $c\alpha$ in the first term of Dirac's famous equation, namely $c\alpha p$, is usually said to stand for the electron velocity, because that term can be interpreted as the kinetic energy. But this must be wrong, because the eigenvalues of α are 1 and -1, whence the electron would always be moving at the speed of light. It took me years to identify the symbol that can be correctly interpreted as the velocity of spin $1/2$ quantons; I also performed a similar job on Kemmer's theory for integral-spin quantons.

Mine were the first axiomatic and realistic formulations of nonrelativistic quantum mechanics, the two relativities, classical electrodynamics, and continuum mechanics. Nearly three decades later the Mexican physicist Guillermo Covarrubias updated my axiomatization of general relativity, while the Argentine physicist Héctor Vucetich and some of his students updated by axiomatization of quantum mechanics. Obviously, most remains to be done or redone.

Most of the philosophers interested in axiomatics prefer the so-called structuralist (or non-statement) view of Suppes, Sneed, Stegmüller, and Moulines. Their work is confined to Newtonian particle mechanics and thermostatics; it overlooks reference frames and frame transformations, it does not include semantic assumptions; and it confuses conceptual models with their physical referents. The late Clifford Truesdell, the greatest authority in classical physics, destroyed this school in 1984. But this news was not allowed to reach the philosophical community: my review of Truesdell's book was rejected.

My work on axiomatics showed me the need to clarify a number of key semantic concepts, such as those of reference (denotation), sense (connotation), and partial truth, as well as of ontological concepts, such as those of system, event, and spacetime. The first two volumes of my Treatise on Basic Philosophy (1974-89) were devoted to semantics (without modal logic, of course). Volumes 3 and 4 of the same work were devoted to ontology, in particular a relational theory of spacetime. Half of volume 7 is devoted to the philosophy of physics.

In conclusion, I regard the physics-philosophy relationship as a mésalliance well worth being formalized. Such normalization should enrich both partners in rigour as well as in problematics.

Selected Bibliography

Ampère, André-Marie. 1834. *Essai sur la philosophie des sciences*. Paris: Bachelier.

Bell, John. 1987. *Speakable and Unspeakable in Quantum Mechanics*. Cambridge: Cambridge University Press.

Beller, Mara. 1999. *Quantum dialogue: The Making of a Revolution*. Chicago: University of Chicago Press.

Bohr, Niels. 1958. *Atomic Physics and Human Knowledge*. New York: Wiley.

Boltzmann, Ludwig. 1979 [1905]. *Populäre Schriften*. Braunschweig: Vieweg.

Born Max. 1971. *The Born-Einstein Letters*. New York: Walker and Company.

Bunge, Mario. 1951. What is chance? *Science & Society* 15: 209–31.

————1959. *Metascientific Queries*. Springfield, IL: Charles C. Thomas.

————1964. Phenomenological theories. In M. Bunge, ed., *The Critical Approach: Essays in Honor of Karl R. Popper*, 234–54. New York: Free Press.

————1967a. *Foundations of Physics*. Berlin-Heidelberg-New York: Springer.

————1967b. Analogy in quantum mechanics: from insight to nonsense. *British Journal for the Philosophy of Science* 18, 265–286.

————1968. Physical time: The objective and relational theory. *Philosophy of Science* 35: 355–88.

————1971. A mathematical theory of the dimensions and units of physical magnitudes. In M. Bunge, ed., *Problems in the Foundations of Physics*, 1-16. New York: Springer.

————1973. *Philosophy of Physics*. Dordrecht, NL: Reidel.

————1979. The Einstein-Bohr debate over quantum mechanics: Who was right about what? *Lecture Notes in Physics* 100: 204–219.

————1982. Is chemistry a branch of physics? *Zeits. f. Allgemeine Wissenschafstheorie* 13: 209–23

————1984. Hidden variables, separability, and realism. *Revista Brasileira de Física* 150–168.

————1987. Treatise on Basic Philosophy, Vol. 7, part 1: *Formal and Physical Sciences*. Dordrecht, NL: Reidel.

————1988. Two faces and three masks of probability. In E. Agazzi, ed., *Probability in the Sciences*, 27–50. Dordrecht, NL: Reidel.

————.2000. Energy: Between physics and metaphysics. *Science and Education* 9: 457–61.

————2003a. Velocity operators and time-energy relations in relativistic quantum mechanics. *International Journal of Theoretical Physics* 42: 135–142.

————2003b. Twenty-five centuries of quantum physics: From Pythagoras to us, and from subjectivism to realism. *Science and Education* 12: 445-66; 587–97.

————2006. *Chasing Reality*. Toronto: University of Toronto Press.

Bunge, Mario, and Andrés J. Kálnay.1983a. Solution to two paradoxes in the quantum theory of unstable systems. *Nuovo Cimento* 77B: 1–9.

————1983b. Real successive measurements on unstable quantum systems take non-vanishing time intervals and do not prevent them from decaying. *Nuovo Cimento* 77B: 10–18.

Campbell. Norman Robert. 1957 [1919] *Foundations of Science: The Philosophy of Theory and Experiment*. New York: Dover.

Covarrubias, Guillermo M. 1993. An axiomatization of general relativity. *International Journal of Theoretical Physics* 32: 2135–54.

D'Espagnat Bernard. 1981. *À la recherche du réel*, 2nd ed. Paris: Gauthier-Villars.

Duhem, Pierre. 1954 [1914] *The Aim and Structure of Physical Theory*. Princeton, NJ: Princeton University Press.

Einstein, Albert. 1950. *Out of my Later Years*. New York: Philosophical Library.

Frank, Philip. 1946. *Foundations of Physics*. Chicago: University of Chicago Press.

Galison, Peter. 1987. *How Experiments End*. Chicago, IL: University of Chicago Press.

Heisenberg, Werner. 1958. *Physics and Philosophy*. New York: Harper & Brothers.

Lenin, V. I. 1950 [1908]. *Materialism and Empiriocriticism*. London: Lawrence & Wishart.

Lévy-Leblond, Jean-Marc. 2006. *De la matière relativiste, quantique, interactive*. Paris: Seuil.

Mach, Ernst. 1910. *Populär-Wissenschaftliche Vorlesungen*, 4th ed. Leipzig: Johann Ambrosius Barth.

Margenau, Henry. 1950. *The Nature of Physical Reality*. New York: McGraw-Hill.

Mehra, Jagdish, ed. 1973. *The Physicist's Conception of Nature*. Dordrecht, NL: Reidel.

Meyerson, Emile. 1925. *La déduction rélativiste*. Paris: Payot.

Pérez-Bergliaffa, S. E., G. Romero, and H, Vucetich. 1993. Axiomatic foundations of non-relativistic quantum mechanics: A realistic approach. *Journal of Theoretical Physics* 32: 1507–25.

Planck, Max. 1933. Where is Science Going? James Murphy, ed. Preface by Albert Einstein. London: George Allen & Unwin.

Truesdell, Clifford. 1984. *An Idiot's Fugitive Essays on Science*. New York: Springer.

4

John Earman

Department of History and Philosophy of Science
University of Pittsburgh, USA

---

1. What is the relationship between philosophy and physics?

There are two ways to read the question. If 'philosophy' (respectively, 'physics') is taken in the narrow sense that is defined by the practices of people who have appointments in philosophy departments (respectively, physics departments), then I think that philosophy of physics is largely parasitic on physics since it focuses on the products of professional physicists.

However, if 'philosophy' is taken in a broader sense that is not constrained by the current disciplinary boundaries, then the presupposition of the question is false because when it comes to basic issues in the foundations of physics, the distinction between philosophy and physics blurs. The activity in this area is perhaps best described by a term that has fallen out of usage—'natural philosophy'. Professional philosophers working on issues in this area will sometimes publish in physics journals; and while the reverse is rarer (I think because leading physics journals are showing an increased willingness to publish articles on foundations issues), it is evident that some of the best work in the area is being done by professional physicists. And at the risk of seeming guilty of self-advertisement, I will mention a concrete example of blurring of the line between philosophy and physics: of the fourteen chapters of the recently published Handbook of Philosophy of Physics, eight where authored by professional philosophers and five were authored by professional physicists.

2. How did philosophers contribute or fail to contribute to the development of physics in the 20th century?

I will assume that 'philosophers' is to be taken in the narrow sense (see 1.). Philosophers can contribute indirectly to the development of physics by influencing the creative processes of theoretical physicists. However, this influence is notoriously difficult to tease out, as witnessed by the debate about what influence Hume's empiricism and Mach's positivism had on the development of Einstein's special theory of relativity.2 The direct contribution of philosophers to the development of physics, either in terms of solving outstanding problems in existing theories or in propounding new theories, is virtually nil. This is true almost by definition. Fellow philosophers would say something like the following about a philosopher who made such a direct contribution: 'He/she is really a physicist, not a philosopher. The proper focus of philosophy is confined to interpretational issues about theories that have been produced by physicists.'

3. What aspect of current work in physics can benefit the most from collaboration with philosophers?

One can give minor examples where such a collaboration would be helpful. For instance, there is a large and growing physics literature on the physical possibility time machines, devices that produce closed timelike curves. One will search this literature in vain for a persuasive analysis (= necessary and sufficient conditions) for the operation of a time machine. Philosophers have pointed out the difficulties in giving such an analysis and have made suggestions for resolving them.3 Awareness of these suggestions could be helpful in the formulation of 'go' or 'no go' theorems for time machines. A more substantial example comes from thermo-statistical physics where there is no settled theory but rather a collection of techniques and competing approaches. The competition often turns on foundations issues, such as the interpretation of the probabilities involved and the source of the temporal asymmetries. These are issues to which philosophers can and have made substantial contributions. As one example, Jos Uffink has shown why the Markov property used in "stochastic dynamics" cannot explain irreversible behavior.4 But for the most part, I am skeptical that physicists who are pushing the frontiers of physics can benefit from collaboration with philosophers. Indeed, most physicists would only be hindered by a knowledge of the philosophy of physics literature or by interaction with philosophers. Understood in this light, philosophers ought not to take offense at Gell-Mann's story that he had a doctor's prescription posted on his wall admonishing him not to debate with philosophers.5 There are, however, notable exceptions. The work of some of the leading researchers in loop quantum gravity (LQG)–specifically Lee Smolin and Carlo Rovelli–seems to benefit from reflection on foundations issues, on which the input of philosophers is welcomed. But these are the exceptions that prove the rule.

4. What area of contemporary philosophy of physics is most fertile?

Work on the foundations of ordinary non-relativistic quantum mechanics (QM) is showing diminishing returns. But there are a host of fascinating foundations issues in relativistic quantum field theory (QFT) that are begging for attention from philosophers of physics. The philosophy of space and time seems largely moribund. But it promises to be revitalized by the study of issues that arise in quantum gravity research. Philosophers who want to work on QFT or quantum gravity must be willing to make a large investment of time and energy to mastering formidable technical material. I am confident that the investment will pay large dividends for the philosophy of physics. A third area that holds exciting prospects for philosophy of physics, but which has been largely neglected by philosophers, is cosmology, which has transformed itself from a largely speculative discipline to a science with ever tighter observational constraints.

5. In your opinion, which area of physics holds the most exciting promise in the coming decades?

My answer to this question echoes my answer to question 4. Combining the insights of quantum physics and Einstein's general theory of relativity (GTR) to produce a quantum theory of gravity is one of the most difficult and exciting challenges facing theoretical physics of the 21th century. As Carlo Rovelli has noted, the current state of physics is reminiscent of pre-Newtonian physics in that we have one set of laws that govern what happens at small scales and another set that governs what happens at large scales. These two sets of laws, if not incommensurable in a Kuhnian sense, are formulated in different languages using disparate concepts. The currently leading approaches to a quantum theory of gravity are superstring theory and LQG, but there are a number of alternative approaches (e.g. those based on Connes's non-commutative geometry and Sorkin's causal sets). Whichever of these approaches triumphs, profound changes in our conceptions of space, time, and matter will surely follow. There will, of course, also be implications for early universe cosmology. But even apart from the early universe regime where quantum gravity effects come into play, cosmology can be expected to generate new discoveries and puzzles, such as the discovery of the accelerating expansion of the universe and the nature of the "dark energy" that is driving the acceleration.

6. How were you initially drawn to the field and what are some examples of your work that have influenced the discipline?

When I was a graduate student (1964-1968) the philosophy of physics as a distinct subdiscipline did not exist. I can't remember how I got interested in the measurement problem in QM. But I do remember being frustrated that I could not drum up any interest in this problem among my fellow students or my teachers (most of whom knew little about QM). Nor was it any good to try to talk to physicists about this problem since at the time they would adamantly deny that there was any problem worth discussing. One exception was Eugene Wigner, whose own solution I did not find attractive—it involved the idea that in an act of measurement, the conscious mind produces change in the physical system being observed. I hit on a version of what is now known as the "modal interpretation" of QM. I worked up my nerve to send Wigner a note outlining my idea. Wigner, who was known for his formidable politeness, sent me a note that was couched in polite language but said, in effect, that one can always solve a problem by ignoring it. I was initially devastated; but eventually I recovered, and I spent a year in Boston where I studied with Abner Shimony. One result was a joint publication on the measurement problem— with scare quotes around 'joint' because Abner wrote the paper and kindly added my name to it. After this I did not work on foundations of QM for many years, partly because I got interested in GTR and partly because I became convinced that the solution of measurement problem required new physics rather than clever interpretational moves.

I do not think that any of my work has any lasting importance. What I would claim on my behalf and on behalf of others of my generation is that we made the philosophical world aware of the exciting vistas philosophy of physics has to offer. And we also trained a new generation of philosophers of physics whose work has already far surpassed any of our contributions.

References

1. J. Butterfield and J. Earman (eds.), Handbook of the Philosophy of Science: Philosophy of Physics. Amsterdam: Elsevier/North Holland, 2007.

2. See J. Norton, "How Hume and Mach Helped Einstein Find Special Relativity," in M. Dickson and M. Domski, eds., Synthesis and the Growth of Knowledge: Essays at the Intersection of History, Philosophy, Science, and Mathematics. Chicago: Open Court, forthcoming.

3. See J. Earman and C. Wηthrich, "Time Machines", Stanford Encyclopedia of Philosophy.
http://plato.stanford.edu/entries/time-machine/

4. J. Uffink, "Compendium of the Foundations of Classical Statistical Mechanics," in Ref. 1.

5. As reported by P. Teller, "The Philosopher Problem," in L. Hoddeson, L. Brown, M. Riordan, and M. Dresden (eds.), The Rise and Fall of the Standard Model: Particle Physics in the 1960s and 1970s, pp. 634-636. Cambridge: Cambridge University Press, 1997.

5

Brigitte Falkenburg

Department of Philosophy
Technische Universität Dortmund, Germany

---

1. What is the relationship between philosophy and physics? What should it be?

In my view, the relationship between philosophy and physics can only be understood against its historical background. Hence, let me start with the question of what it was in the past. Physics emerged from philosophy. In 17th century, physics was actually called "natural philosophy" (Newton) or "experimental philosophy" (Locke). Indeed, the conceptual foundations of classical mechanics emerged from 17th century metaphysics, in particular, Neo-platonism and rationalism. Newton's concepts of absolute space and time did not only define an ideal frame for the motions of mechanical bodies. His concept of space was also rooted in the Neo-Platonist conviction of God's omnipresence in the world through an immaterial substance. This metaphysical conviction was not shared by his opponent Leibniz, who believed in the pre-established harmony of infinitely many immaterial monads. The general concepts of substance and cause, however, were common to all metaphysical positions, even though every philosopher believed in other kinds of substances and causal agents. In physics, the concept of substance gave rise to the concept of an independent physical object, such as a mechanical body with an inertial mass. The concept of cause, in turn, gave rise to the idea of a physical dynamics based on differential equations and their solutions. The causal agents of physics, i.e., forces, fields, and atoms, were conceived as independent substance-like entities, too.

The relationship between philosophy and physics was mutual, but in the other direction it did not work. Philosophers from Descartes to Kant wanted to establish the principles of metaphysics by modeling their methodology after physics or mathematics. In order to do so, they used the so-called analytic-synthetic or resolutive-compositive method of mathematics or physics. In Galileo's experimental method, Descartes' analytical geometry, Newton's mechanics and optical experiments, Leibniz' logic and mathematic it was most successful. In Descartes' metaphysical dualism, Spinoza's ethics more geometrico, or Leibniz' monadology, however, it was not. The search for unquestionable metaphysical principles was in vain. It gave rise to diverging concepts, contrary principles and methodological confusion. Kant reacted to these fruitless attempts (including his own pre-critical metaphysics) by developing his critical philosophy. His critical theory of nature was based on the conviction that the conceptual foundations of our knowledge are subjective. However, this did not mean that our concepts of space and time or the principles of substance and cause are arbitrary. Kant thought that they are indispensable transcendental conditions a priori of possible experience. On this basis, he came to believe that the foundations of Newton's physics are the only remaining metaphysical knowledge of nature.

Starting with Kant, the exact sciences and natural philosophy definitely split. 19th century physics proceeded without any need of applying Kant's Metaphysical Foundations of Natural Science to the concepts and principles of thermodynamics, electrodynamics, or kinetic theory. Schelling's speculations about the forces and polarities in nature were stimulating for J.R.Mayer's discovery of energy conservation and H.C.Oerstedt's research on magnetism. But then, the gap between German idealism (in particular, Hegel's philosophy) and the empirical sciences became enormous. Now, the triumphal 19th century march of positivism and empiricism began. It culminated in Mach's philosophy, which was most influential around 1900, competing with the idealistic Neo-Kantian views about the exact sciences and the humanities. Mach criticized Newton's concepts of absolute space, absolute time, and force as metaphysical concepts without any empirical content. He proposed to reformulate classical mechanics without them. In addition, he criticized atomism and the kinetic theory of Boltzmann, his follower at the University of Vienna. The first criticism was very inspiring for 20th century physics, the second obviously not.

From an empiricist point of view, physics deals with empirical substances and processes only: macroscopic phenomena and their observable properties; operational concepts based on well-defined measurement methods; empirical laws from measurements and phenomenological laws expressed in operational terms. In addition, empiricism tends to a Humean regularity view of causality, according to which there are no necessary laws of nature. For Mach, the unifying laws of mathematical physics only express the economy of human thought. This view was in obvious conflict with the rationalist ideas behind modern physics. In an influential talk about the unity of physical reality given in 1908, Planck sharply criticized Mach's views, in defense of scientific realism. He argued that the unobservable forces and structures of physics express a true unified mathematical reality that underlies the variety of the phenomena.

For the development of physics, strict empiricism indeed was sterile. It was more fertile to combine some ideas of empiricism with other philosophical principles. Hermann von Helmholtz showed this with respect to Kant's concepts of space and time. In 20th century physics, the tradition of an epistemology influenced by empiricism but not restricted to it went on. Einstein had a hybrid epistemology which was a mixture of rationalist and empiricist ideas. Bohr's and Heisenberg's Copenhagen interpretation of quantum mechanics, in turn, is based on a Kantian quest for intuitive concepts plus Mach's view that scientific explanation rests on analogies.

So, what does history teach us about the relationship between physics and philosophy? Galileo's resolutive-compositive method aimed at analyzing the phenomena in order to find the underlying causal agents and to describe them in mathematical terms. But the conceptual and epistemological frame of modern physics was rationalism. The general metaphysical concepts of substance and cause were most fertile for the development of classical physics. But when metaphysics in turn copied the successful methods of physics or mathematics, it went astray. Kant's transcendental philosophy attempted to keep the foundations of Newtonian physics as a reliable part of metaphysical knowledge. But this attempt failed, too.

Mach's empiricism helped to criticize metaphysical concepts without empirical content, above all, Newton's concept of absolute space. But his critical attitude towards atomism shows that too strict empiricist views hinder rather than inspire the development of physics. 20th century physics was epistemologically less strict and conceptually more successful in renewing the foundations of physics.

The question of how the relation between philosophy and physics should be remains. This is easy to answer. It should be stimulating for both disciplines. Philosophy should provide physics with new insights into our knowledge of nature, and vice versa. Traditional philosophy inspired physics with a mixture of metaphysical speculations and epistemological criticism. 20th century physics, in turn, challenged philosophy to interpret and comment its theories.

2. How did philosophers contribute or fail to contribute to 20th century physics?

How did the above story continue? Mach's philosophical analysis of Newton's mechanics was straightforward and convincing, but his empiricist quest for strictly phenomenological theories did not contribute to the scientific revolutions of 20th century physics. Relativity and quantum theory only became possible when the physicists learned how to combine their own, genuine scientific realism with selected elements of Mach's empiricist methodology.

Einstein trusted in operational concepts as far as they agreed with his quest for theoretical unification and axiomatic theories. This was the case for his operational concept of simultaneity that gave rise to Special Relativity, but not for the operational concepts of quantum mechanics. Einstein objected against Heisenberg's matrix mechanics that the theory should tell what can be measured, and not vice versa (as Heisenberg reported in his Physics and Philosophy).

Bohr, on the contrary, weakened the traditional goal of giving a unified theoretical explanation, which was pursued by Newton, Maxwell, and Einstein. For Bohr, a scientific explanation was a classification of the phenomena in terms of analogies. His quantum postulates were based on analogies between quantum concepts and classical concepts supported by his correspondence principle. Old quantum theory combined quantum laws and classical laws in such a way that Hilbert considered it to be far from admitting of axiomatic foundations. Old quantum theory, however, failed for the anomalous Zeeman effect of the spectra of complex atoms. Finally, Heisenberg eliminated the unobservable electron orbits from quantum theory. In his matrix mechanics, he kept only operational concepts, namely the energy, frequency, polarization, and intensity of the observable spectral lines. This was what Einstein criticized from his axiomatic point of view. Schrödinger's wave mechanics was more complete. It replaced the unobservable electron orbits (which Heisenberg had completely skipped) by uninterpreted wave functions. Therefore, Born suggested the probabilistic interpretation of quantum mechanics. This, in turn, could only give an indirect operational interpretation to Schrödinger's wave function.

In all of the above cases, the physicists criticized non-operational concepts that corresponded to unobservable metaphysical entities. But they only did so in order to keep other crucial elements of an undoubted scientific realism. This is most obvious in Einstein's case. Einstein believed in unified laws of nature and wanted to express them in terms of axiomatic theories. All of his work on special and general relativity, the light quantum hypothesis, and Brownian motion was in search of unification. In order to unify, he criticized Newton's metaphysical concepts of absolute space, time, and simultaneity, on the one hand, and pushed atomism forward, on the other hand. But when he succeeded to explain Planck's law of black-body radiation from Bohr's atomic model in terms of his theory of the absorption and emission, he was quite uneasy about the non-deterministic behavior of the light quanta. Finally, in view of quantum mechanics, he was not willing give up his quest for a complete description of physical reality in terms of objects with well-defined spatio-temporal and dynamic properties.

The founders of quantum mechanics and the Copenhagen interpretation, too, kept their belief in atomism. Bohr and Heisenberg were willing to give up the unity of physics and the concept of an object with unambiguous physical properties, but they were not willing to give up their belief in the existence of subatomic reality. Consequently, they were very far from being happy when the founders of logical empiricism considered their interpretation of quantum mechanics to be grist on empiricist mills.

The most important contributions of the philosophers to 20th century physics were indirect and unintended. They stemmed from philosophical reading and philosophical undergraduate studies rather than direct influence or collaboration. Finally, Bohr and Einstein developed their own philosophies, which gave rise to the Bohr-Einstein debate on the interpretation of quantum mechanics. This debate definitely was the most important contribution of 20th century philosophy to the development of physics. It opened the whole field of non-local quantum phenomena in quantum optics, condensed matter physics, and other domains, and even gave rise to new technological ideas such as quantum cryptography and quantum computation.

Indeed, physics made more substantial contributions to 20th century academic philosophy than vice versa. Without Einstein's theories of relativity and the quantum revolution, Carnap and Reichenbach had probably seen no reasons to turn from Neo-Kantianism to Mach's empiricism. In such a counterfactual world, logical empiricism had not become so influential.

But this is not yet the whole story. Here, Carnap's and Reichenbach's Neo-Kantian background comes into play. Around 1900, Neo-Kantianism (in particular, the Marburg school founded by Hermann Cohen) had criticized Kant's view of space and time. To be more precise, they re-interpreted Kant's forms of intuition in terms of logical concepts. This collapsed Kantianism into some version of Neo-Platonism and went together with constructivist views about physics. Concerning the empirical basis of physical theories, Cohen's views tended to a version of idealism that was close to the opposing empiricist views of Mach's. Cohen emphasized the theory-dependence of empirical data in a way that gave rise to anti-realism and instrumentalism about physical theories. So, in view of relativity and quantum theory Carnap and Reichenbach were inclined to positivism and, at the same time, to identify Kant's theory of nature with the bad metaphysics that had failed.

And this gave rise to a most important regard in which the philosophers failed to contribute in a stimulating way to 20th century physics. Their increasing esteem of empiricism and disesteem of Kant's philosophy brought Carnap and Reichenbach to radical verificationism. On this basis, Carnap initiated an influential research program, namely the search for well-defined criteria for the demarcation of empirical science from metaphysics. Later, the Duhem-Quine thesis and other developments in the philosophy of physics showed that this search was in vain. (Indeed, the search for such demarcation criteria was no less in vain than the fruitless attempts of 17th and 18th century metaphysics to copy the analytic-synthetic method of mathematics or physics in order to find reliable metaphysical principles.) It turned out that there is no physics without metaphysics. However, this vain research and its inner-philosophical criticism were not really inspiring for the few physicists who continued to be interested in philosophy, in the post-war era.

To these unproductive misunderstandings of 20th century physics, the contingent facts of 20th century German history added. As mentioned above, it was not Kant's critical philosophy itself that was refuted by 20th century physics, but rather its Neo-Kantian version on the one hand, and the pre-Kantian metaphysical concepts of substance and causality, on the other hand. The only Neo-Kantian philosopher who spelled the latter point out was Ernst Cassirer. But he left the Third Reich to Sweden, whereas Carnap and Reichenbach went to the USA. Hence, there was no more philosophical discussion between the outstanding 20th century Neo-Kantian and the most influential 20th century empiricist philosophers of science.

However, Carnap's search for sharp demarcation criteria initiated another more fertile research program, namely the reconstruction of the structure of physical theories and their empirical basis and substructures. This research program had much offspring. The most outstanding empiricist philosophers of science contributed to it. In addition, several physicists were interested in it. In Germany, it influenced G.Ludwig's approach to the axiomatic foundations of quantum theory. And within this research program, P.Suppes and his collaborators developed the modern theory of measurement, which became a branch of applied mathematics.

A second important impact of philosophy on physics concerned Th.S.Kuhn's insight into the incommensurability of the theoretical concepts before and after scientific revolutions. The postmodern tendencies of social constructivism and cultural studies after Kuhn surely were far from giving any contributing to physics. But Kuhn's famous book itself impressed the physicists a lot. According to my teaching experience, it is attractive for physics students up to the present day. Even R.P.Feynman (whom no one could suspect of having been affected too much by professional philosophy of science) gave the following Kuhnian account of the change in the meaning of mass in the transition to relativistic mechanics, in the Feynman lectures:

"Philosophically we are completely wrong with the approximate law. Our entire picture of the world has to be altered even though the mass changes only a little bit. [...] Even a small effect sometimes requires profound changes in our ideas."

3. What aspect of current work in physics can benefit most from collaboration with philosophers?

As emphasized above, philosophy may stimulate physics with a good mixture of metaphysical speculations and epistemological criticism. In addition, physicists may profit from studying the philosopher's insights into the structure of scientific theories, their empirical foundations, and their metaphysical background. The latter, however, was neglected in 20th century philosophy of science.

To be more specific, the heterogeneous structure of current physics has to be mentioned. Physics is far from having a unified axiomatic basis. It consists in a mixture of non-relativistic and relativistic, classical and quantum, particle and field theories, phenomenological models, and measurement theories. The parts of this wild architectonics are connected by tacit use of inter-theoretic relations (such as Bohr's correspondence principle or its generalization), heuristic rules (e.g., dimensional arguments), symmetry principles and conservation laws, formal analogies, and the extrapolation of the scales of physical quantities from the Planck scale to the size of the universe. To these elements of theoretical physics add measurement methods, experimental devices, the technologies involved them, the mathematical methods of data analysis (e.g., statistical methods and the estimation of errors), computer simulations, diagrams and pictures that illustrate crucial model assumptions, and so on.

There are two main reasons for the heterogeneity of current physics. One is the complexity and non-linear behavior of many physical systems, the other the unresolved quantum measurement problem. In this situation, the collaboration of physics and philosophy is important in several fields: (i) for conceptual unification, above all, concerning the conceptual problems of quantum gravity; (ii) for a better understanding of the "fuzzy" classical-quantum borderline, including the successes and limitations of decoherence; (iii) for methodological, structural, conceptual, and epistemological insights into the weird architectonics of current physics; (iv) perhaps, also for the complex problems of data analysis and model building in experimental physics. An interesting case of collaboration in astrophysics, where philosophers accompanied the data analysis and model building of the physicists, is due to G.Grasshoff (Bern).

4. What area in contemporary philosophy of physics is most fertile?

During the second half of the 20th century, physics and the philosophy of physics split into separated scientific communities. But philosophy of physics will only become fertile when it supports the interplay of both disciplines. In order to make this possible, the philosophers should relate their work to the current research programs of physics. Concerning quantum field theory, particle physics, cosmology, and the search for quantum gravity, this is the case. Concerning the other exciting fields of current physics, in particular, quantum optics, condensed matter physics (i.e., the physics of complex systems), and astroparticle physics, the philosophers still are too reserved. Here, more philosophical work (such as M.Morrison's) on the conceptual problems of unification, model building, and the limitations of reduction is desirable. Such work would result in learning more about physical practice and, hopefully, about the quantum-classical limit, too.

In addition, current philosophy should try to have more metaphysical and epistemological impact on physics. In my view, the metaphysical and epistemological lesson of quantum theory has not yet been learnt. The quantum measurement problem has remained unresolved for 80 years. The most advanced approach, namely decoherence, is only at the probabilistic level able to resolve it. Concerning the objective outcome of an individual quantum measurement, there are several no go theorems, and there is no idea of a physical mechanism of objectivation.

In this situation, Bohm's hidden variables and the many-world interpretation of quantum theory have drawn much attention. To my opinion, however, they are in need of philosophical criticism. I have the following objections against them. They are ad hoc solutions for the measurement problem that remain sterile for the rest of physics. It may be countered that they make quantum cosmology possible. But, they generate more conceptual problems than they intend to resolve. A Bohmian approach obviously does so for relativistic quantum field theory. The many-world interpretation does so for the modal categories of the possible and the actual. (What does it mean that all possible measurement outcomes become actual, in splitting worlds?) Furthermore, they multiply the ontology of physics without any hope of measuring the additional structures. This empiricist point should not be ignored. After all, physics should remain an empirical science.

I suspect that both kinds of speculation miss the metaphysical and epistemological lesson of quantum theory. Subatomic structure forced the physicists to dispense with the concept of a physical reality made up of classical particles. Wave-particle duality was no arbitrary invention but the only viable way of coping with the empirical structure of matter and light, that is, with the photo effect, Rutherford scattering, the spectra of complex atoms, the Franck-Hertz experiment, electron interference, and so on. Classical particles (or fields) with complete spatio-temporal and dynamic properties are conceived as independent causal agents in the sense of the traditional metaphysics. Hence, the subatomic structure of physical reality disagrees with the metaphysics of classical physics. Neither empiricism nor any attempt to re-establish metaphysical realism in terms of substance-like entities (be they Bohmian particles or a block universe) will make the quantum structures more comprehensible which show up in a classical world.

For all these reasons, I hope that the current revival of Bohr's complementarity philosophy and of Neo-Kantian approaches to the metaphysical foundations of quantum physics will become fertile. Important work in this direction is due to M.Bitbol (CREA, Paris), P.Mittelstaedt (Cologne), H.Pringe (Barcelona and Dortmund), H.Zinkernagel (Granada). Indeed, Bohr's complementarity view of quantum mechanics is closely related to crucial ideas of Kant's philosophy.

5. In your opinion, which area of physics holds the most exciting promise in the coming decades?

Concerning the offspring of the current theoretical research programs of superstrings, quantum loop gravity, or other approaches to the physics at the Planck scale, I am a bit pessimistic. Their main problem is the lack of possible empirical support. There are no experiments to test them directly. In addition, I can not see how they may resolve the quantum measurement problem, which is a main obstacle for the attempts to unify quantum theory and cosmology.

If the LHC experiments at CERN will find supersymmetric particles, the Higgs boson, or more, the standard model of current particle physics will be seen in new light. If the experiments of astro-particle physics identify the sources of high-energetic cosmic rays, the nature of dark matter, or more, this will be even more exciting. Another most important discovery would be the detection of gravitational waves. Hence, the most exciting promises of the coming decades are new measurement results in the intersection of particle physics, astrophysics, and cosmology.

How were you initially drawn in the field and what are some examples of your work that have influenced the discipline?

My background is twofold, experimental particle physics and traditional philosophy of nature. After finishing my diploma in atomic physics (TU Berlin), I studied traditional philosophy, with the focus on German idealism (University of Bielefeld). What followed was an adventurous zigzag path between physics and philosophy, culminating in parallel work on two PhD's, one on Hegel's and Kant's philosophy of space, time, and matter (Bielefeld 1985), the other in experimental high energy physics (Heidelberg 1986: measurement of nucleon structure functions in deep-inelastic neutrino-nucleon scattering at CERN). Afterwards, I worked with Erhard Scheibe (University of Heidelberg). There, I wrote my habilitation thesis on the experimental basis and the conceptual problems of particle physics. In this way, I changed from particle physics to particle metaphysics. In the academic year 1995/96, I was a fellow at the Institute for Advanced Study in Berlin. There, I participated in the project Models in Physics and Economics, with M.Morrison and M.Morgan. (An offspring of this project is my recent work on the relationship between physics, economics, and technology, including the problems of climate change.) Since 1997, I have the chair for philosophy of science and technology at the TU Dortmund.

My work in the philosophy of physics is centered on the empirical and conceptual foundations of particle physics, Kant's theory of nature, Niels Bohr's philosophy, the correspondence principle, wave-particle duality, and the limitations of reduction in physics.

My recent Particle Metaphysics investigates the experimental basis and the metaphysical import of the particle concept, from the discovery of the electron and photon to the quark structure of the nucleon and the which way experiments of quantum optics. In particular, I analyze the theory-ladenness of the data (particle tracks, scattering events, resonances), the causal and mereological aspects of the particle concept, the hidden unity behind the heterogeneous measurement theory of particle physics, the current family of particle concepts (including field quanta, Wigner's group theoretical definition, virtual particles, and the quasi-particles of solid state physics), and the conceptual history of wave-particle duality from the beginnings to recent quantum optics. With these investigations, I attempt to shed new light on the debate on scientific realism.

My book Kants Kosmologie investigates the relationship between philosophy and physics in Kant's cosmology, with emphasis on Kant's pre-critical use of the "analytic-synthetic" method, the way in which this method failed with regard to the concept of space, the emergence of Kant's critical philosophy from this problem, and a detailed analysis of Kant's "antinomy of pure reason" and its relation to 20th century physics. These investigations are the background of my attempts to apply Kantian ideas to quantum physics, in my Particle Metaphysics and in a series of papers.

Recent Books:

Particle Metaphysics. A Critical Account of Subatomic Reality. Heidelberg: Springer 2007.

Wem dient die Technik? Eine wissenschaftstheoretische Analyse der Ambivalenzen technischen Fortschritts. Prize Essay In: Die Technik – eine Dienerin der gesellschaftlichen Entwicklung? Johann Joachim Becher-Preis 2002 (ed.: J.J.Becher-Stiftung Speyer), Baden-Baden: Nomos 2004, 45-177.

Kants Kosmologie. Die wissenschaftliche Revolution Naturphilosophie im 18. Jahrhundert. Frankfurt am Main: Klostermann 2000.

Selection of articles:

A Critical Account of Physical Reality. To appear in: M.Bitbol (ed.), Constituting Objectivity. Dordrecht: Springer.

The Invisible Hand: What Can we Know? In: Epistemology and the Social. E.Agazzi, J.Echeverría, A.Gomez (eds.), Proceedings of the Colloquium of the International Academy of Philosophy of Science (Tenerife 2005), to appear.

Functions of Intuition in Quantum Physics. In: Intuition and the Axiomatic Method, ed. by E.Carson and R.Huber, Western Ontario Series in the Philosophy of Science, Dordrecht: Springer 2006, 267-292.

Intuition and Cosmology: The Puzzle of Incongruent Counterparts. In: Intuition and the Axiomatic Method, ed. by E.Carson and R.Huber, Western Ontario Series in the Philosophy of Science, Dordrecht: Springer 2006, 157-180.

Some Remarks on Cosmology and Scientific Realism. In: Kairos 26 (2005), 229-246.

Experience and Completeness in Physical Knowledge: Variations on a Kantian Theme. In: Philosophiegeschichte und Logische Analyse, Schwerpunkt: Geschichte der Naturphilosophie (Hrsg.: U.Meixner und A.Newen). Paderborn: Mentis 2004, 153-176.

Kant's Architectonic Principles for a Metaphysics of Nature. In: The Kantian Legacy (1804-2004): Unsolved Problems (ed.: Cinzia Ferrini). Napoli: Bibliopolis 2004, 127-153.

Symbol and Intuition in Modern Physics. In: M.Ferrari and I.-O. Stamatescu (eds.): Symbol and Physical Knowledge. Heidelberg, Berlin: Springer 2002, 149-176.

Measurement and ontology: What kind of evidence can we have for quantum fields? In: Ontological Aspects of Quantum Field Theory, eds. H.Lyre, M.Kuhlmann and A.Wayne, Singapore: World Scientific 2002, 235-254.

Bohr's Principles of Unifying Quantum Disunities. In: Philosophia naturalis 35 (1998), 95-120

Incommensurability and Measurement. In: Theoria, Vol. 12 Numero 30 (1997), 467-491.

6

Steven French

Division of History and Philosophy of Science

Department of Philosophy

University of Leeds, UK

The Entangled Relations Between Physics and Philosophy

The nature of the relationship between philosophy and physics varies, depending on the sub-cultures of philosophers and physicists. There are metaphysicians who purport to tell us how the world is in all its metaphysical finery but who either ignore modern physics entirely, drawing on knowledge of everyday items or a dimly remembered understanding of classical physics at best, or who attempt to invoke quantum phenomena but fall back on semi-classical descriptions which undermine the relevance of their conclusions. There are physicists who similarly claim to tell us how the world is, as viewed through the dark glasses of our latest theories, but hang the resultant picture on philosophically naïve understandings of space, time, objects, identity and so on, and thus fail to do justice to the significance of their own work. We can, of course, do better than this and put the relationship between the two on a more productive footing.

Let me slip into biographical mode for a second: While completing my undergraduate degree in physics at the University of Newcastle upon Tyne, I became interested in foundational issues. Although the department was heavily oriented toward practical applications, particularly in the field of geophysics, I was fortunate in having a physics tutor who appreciated the significance and value of foundational studies. He suggested I speak to a colleague in the philosophy department, who in turn directed me to Chelsea College, University of London, principally because of the work of their new lecturer in philosophy of physics, Michael

Redhead. The history and philosophy of science department at Chelsea was unique in a number of ways, not least because they only accepted postgraduates with a science or maths background. I was fortunate to be accepted for the PhD programme and spent the first year taking a number of (crash) courses in philosophy, history and philosophy of science and logic. My thesis topic developed from an essay I wrote for Professor Heinz Post but it was Michael who became my supervisor and not only directed me to focus on the significant issues but also showed me what it was to be a philosopher of physics.

That topic had to do with issues of identity and individuality of quantum objects and can be approached by thinking about the following question: can we treat quantum objects as individuals, with well-defined identity conditions, just like rocks, tables, cats and people? The standard view for many years, originating with some of the quantum revolutionaries themselves, like Heisenberg and Born, was that no, we can't, that quantum objects are non-individuals in some not entirely clear sense that was typically articulated via metaphor: so quantum particles were portrayed as like pounds or dollars in the bank – you could say how many you had, but you couldn't pick any out and say 'they're mine'. It turns out we can go beyond such metaphors and represent such a view in terms of non-standard set theories (I was introduced to these by my Brazilian colleagues and collaborators, Newton da Costa and Décio Krause; and just to give a plug: together Décio and I have expressed our approach in Identity in Physics: A Formal, Historical and Philosophical Account, OUP 2006).

But also, a number of us – 'us' being philosophers of physics – argued that one could give a positive answer to this question and treat quantum objects as individuals, albeit subject to certain constraints on what states are accessible to them. The implication, then, is that the physics itself does not determine which metaphysical interpretation we should adopt – particles as individuals or particles as non-individuals - a claim which itself sheds some light on this over-arching question of the relationship between physics and philosophy.

Now, if we can regard quantum objects as individuals, then the further philosophical question arises: what grounds this individuality? Do we take it as primitive? Or as having something to do with some underlying 'bare' substratum? Or does it have something to do with the full set of properties of the objects? Taking this last option, for example, Leibniz's famous Principle of the

Identity of Indiscernibles basically acts as a guarantor for individuality by insisting that no two objects can possess exactly the same set of properties; so one can always find some discernible difference between two objects by which they can be individuated. Philosophers have whiled away many a wet Sunday afternoon pondering this Principle, dreaming up spartan possible worlds containing two indiscernible iron globes, for example, and concluding that the Principle could hence not be necessary, but must surely still be contingently true since ... well, their reasons to be inserted in the ellipsis are little better than Leibniz's own command to Princess Caroline that she scurry about the Hohenzollern gardens and just try to find two leaves exactly alike.

Again, some of 'us' philosophers of physics have considered this issue of the status of Leibniz's Principle, not by drawing on our experience of everyday objects like leaves and such, but in the context of modern physics. And some (further sub-set) of us have concluded that on a standard interpretation of quantum mechanics, the Principle is violated (French and Redhead). Others have resisted this conclusion, arguing that a form of the Principle is still viable (Saunders); but the point is that these arguments are informed, not just philosophically, of course, but by an appropriate understanding of the relevant physics. That is how metaphysical discussions should proceed.

But we shouldn't let the physicists of the hook! One of my most vivid memories from an early stage of my career is of a conference of historians, (some) philosopher and physicists, at which I tried to argue that quantum objects could be treated as individuals, but that their individuality had to be considered as 'transcending' their properties. Afterwards a very famous physicist (who shall remain nameless) took me to one side and in a very patronizing manner, explained that individuality was really a very simple matter – and at this point a stooge appeared, holding two ping pong balls – consisting, in fact, of noting that one object (ping pong ball) was 'here', and the other 'there', and that if we interchanged them, for example, we could follow their movement and, really, that was all there was to it. All the talk of 'primitive thisness' or Identity of Indiscernibles was just so much philosophical mumbo-jumbo!

At that point I all but despaired. Here's another memory, from many years later, of a conference which, bravely but perhaps ambitiously, tried to bring philosophers, physicists and historians of science together to discuss the foundations of quantum field theory, where an array of august Nobel prize winners sat up on the stage and treated questions from the philosophical segment of the audience with barely disguised contempt and shot down with withering scorn even those who had expended considerable time and energy trying to understand the details of the theories under discussion. It was so bad that one of the younger physicists was moved to rebuke his colleagues for their bad behaviour.

It wasn't always so, of course. As I mentioned above, many of those involved in the quantum and space-time revolutions of the early twentieth century were not only philosophically informed, in the sense of having at least some education in the subject, but felt no qualms in drawing out the philosophical implications of their work. Some of course were more philosophically 'aware' than others: consider Fritz London, one of the first post-revolution 'normal' scientists (in Kuhn's sense) and a clear contender for 'one of the greatest physicists who never won a Nobel' award. He is perhaps most famous for his quantum theory of the homopolar bond with Heitler and his work on superfluidity, but he began academic life as a philosopher and a phenomenological one to boot! His early philosophical thinking, influenced strongly by Husserlian phenomenology, had a direct impact on both his understanding of the foundations of physics and his view of how theories should be constructed and developed. Indeed, the hugely influential London and Bauer monograph on the measurement problem is written from a phenomenological perspective, so that what people have always taken to be the introduction of consciousness into the quantum realm is of course nothing like; rather, the so-called 'collapse' of the wave-function and the creation of the relevant belief state are seen as co-emerging objective and subjective 'poles' respectively. This is perhaps an extreme example but if philosophers (and physicists) had only paid more attention, the way discussions of the measurement problem developed may have been very different. And the well-known model of superconductivity he elaborated with his brother provides a nice methodological example in which theory construction proceeds through a two-stage process: first one presents an adequate formulation of the relevant phenomenon represented by what he called a 'macroscopic' model, and then one proceeds to a theoretical and, ultimately, microscopic, explanatory schema (see K. Gavroglu, Fritz London: A Scientific Biography, CUP 1995).

Of course, the traffic went the other way as well, with Cassirer, Reichenbach, Russell and others taking hold of the new physics and setting it in new philosophical frameworks. Cassirer's neglected Determinism and Indeterminism in Quantum Physics, for example, takes a neo-Kantian line and argues that the true import of quantum mechanics has to do not with determinism and causality but with the classical notion of object, which, putting it bluntly, has to be dropped from our metaphysics. Likewise, Russell's Analysis of Matter needs to be put in its historical context as a smudged glass through which the emerging quantum physics can be dimly viewed. There are also clear examples in which certain notable philosophers contributed to the development of physics both by exposing and elucidating the conceptual foundations of theories and by helping to set the direction in which developments should proceed. The explosion of work on the foundations of quantum mechanics that followed the work of Bell is an obvious example. Less well-known perhaps is the kind of contribution mapped out so beautifully by Ryckman in his wonderful book, The Reign of Relativity (OUP, 2006). Ryckman shows how the kind of Husserlian phenomenology that influenced London also permeated the work of Weyl, for example, and made a huge contribution to the latter's conception of space-time, the development of the gauge principle and his overall understanding of General Relativity. Similar influences can be traced running through Eddington's work, both at the foundational and more generally theoretical levels. Of course, some might say that it was precisely such influences that led the likes of Eddington astray and contributed to the impenetrability of his last work, Fundamental Theory (CUP 1946).

But perhaps that's just how things look retrospectively through positivist spectacles. At least Eddington felt the need for philosophy's contribution, whereas post-Copenhagen physics has kept it at bay. Of course, Heisenberg's mis-understanding of Einstein on observability is a well-known example of the contribution of broadly positivistic philosophy to the direct development of a piece of physics, which in a nice loop of feedback was itself drawn upon by the (later) positivists to cement into place not only their view of the foundations of physics (see for example, the likes of Carnap and Bridgman on the diminished role of models in quantum theory) but their understanding of how physics should be done. The likes of London and Eddington are atypical of course. Generally physicists remain immune or even hostile to methodological suggestions from philosophy. Whether you think that is as it should be may depend on your view - normative or descriptive

- of the relationship between science and some broader context, whether that be social, political, economic or philosophical. It is perhaps even more a cause for dismay that they have also remained similarly reluctant to draw on philosophical resources in the articulation and resolution of foundational issues, although as I have already indicated, this may just be a post-war 'blip' in the physics-philosophy relationship.

But then something happened, There's probably some neat sociological explanation having to do with the shift in the centre of gravity of physics from Europe to the USA, and the consequent change in cultural, educational and perhaps even political context, with the rise of logical positivism and perhaps even with the need to stop philosophical reflection and get on with war work. Whatever the case, post-war physicists seemed not only uninterested in philosophy, but actively scornful of it. On the philosophy side, a thriving community of philosophers of physics did emerge, at least, with many publishing in physics journals and interacting with those physicists who were similarly interested in foundational issues. Nevertheless, the majority of mainstream metaphysicians at best pay only lip service to the physics, and their counterparts in physics pay even less, often lumping philosophers of physics in with scientifically aware sociologists, sociologically oriented historians and other critics (and indeed, the latter have rebuked philosophers of physics and of science in general for their political impotence, although one can surely contribute to an understanding of foundations while remaining critical about the allocation of resources and selection of aims).

Lest I paint too gloomy a personal picture, let me say that there are stand-out and stand-up exceptions on both sides, from the young physicist I mentioned above, to certain metaphysicians who do indeed bring a finely tuned appreciation of fundamental physics to their philosophical work. And of course the relationship between physics and philosophy should not be shaped and twisted by ignorance and suspicion. Not should it be one of master and servant, where philosophers serve as physicists' handmaidens, there only to re-present the core ideas and notions and views in simpler terms to the great mass of lay-folk. Rather, the relationship should be one of constantly evolving interaction, with philosophers tackling these core ideas and notions, putting them in an appropriate metaphysical context and subjecting them to analysis and criticism, relating them to other concepts and terms and re-presenting them back to the physicists. Who should in turn take the trouble to learn some philosophy (its not that hard!) and get to grips with the relevant metaphysics, critiquing it in their turn, suggesting ways it might be extended and further developed. And within such over-arching reciprocity, philosophy might do more than simply clear away the undergrowth from around theories, exposing both their foundations and inter-connections and contribute to the building and development of new structures. In the process, philosophy itself would become both more relevant and, insofar as such terms can be applied to philosophical views, more correct or true, based as it would be on our best theories of how the physical world is, or could be, rather than some outdated, semi-classical armchair understanding.

We can perhaps discern the potential benefits of a developed form of collaboration between physics and philosophy in the foundations of quantum gravity, where not only the physics, but the metaphysics of quantum mechanics and general relativity meet. And where physicists and philosophers can themselves meet up and work together on foundational issues. The way forward has already been indicated by Huggett and Callendar's well-known collection, Physics Meets Philosophy at the Planck Scale, but let me tie my answer to a personal project which resulted in The Structural Foundations of Quantum Gravity, OUP 2006, co-edited with Dean Rickles and Juha Saatsi. This originally resulted from a series of discussions here at Leeds on structuralist approaches to physics, centred around a series of papers which, on the one hand, re-examined the notion of object in quantum physics and, on the other, re-considered the nature of the underlying manifold in space-time theory.

With regard to the former, the basic idea is to take seriously the concerns touched on above, about identity and individuality of quantum objects and urge a 'reconceptualisation' of such objects as nodes in a structure. This forms part of an attempt to construct a form of scientific realism appropriate for the quantum realm that was originally developed by a PhD student here at Leeds, James Ladyman. The view is known as 'ontic' structural realism because it proposes that quantum objects should be ontologically reconceptualised in structural terms.

This in itself has provoked a range of metaphysical issues to do with the notion of structure, how that structure, or structures, should be seen from the perspective of quantum theory, how this position relates to similar structuralist views of mathematics (the last being examined in particular at a recent conference at the

University of Bristol which brought philosophers of physics and philosophers of mathematics together) and so on. James and I have tried to bring considerations from the foundations of physics into a suitable metaphysical framework in terms of which a suitable notion of structure can be elaborated that is modal and multi-aspected and capable of accommodating the quantum features of the world. The physicists' reduction of particles to sets of invariants is straightforwardly represented within this form of structural realism and the role of group theory can thus be accommodated. Reflecting on this role has led me to look back to the introduction of group theory into quantum mechanics through the work of Weyl, Wigner and von Neumann in the late 1920s. This powered not only important developments in the mathematical framework of quantum mechanics but also with regard to philosophical reflection on the theory, particularly, of course, structuralist reflection. The history here has yet to be told properly and it can be characterized as a wonderful intertwining of physics and philosophy. Typically when structural realists look back to the antecedents of this view they focus on Russell and Poincaré. Russell's The Analysis of Matter, mentioned above, was written in 1926 and published in 1927 and as I said, the still emerging quantum revolution can be discerned only dimly in its pages. Unfortunately this work has cast a long and distorting shadow over subsequent and metaphysically more interesting structuralist developments in the post-revolutionary world. These developments can be traced through the work not only of Eddington and Cassirer but also of Born and Schrödinger and the manner in which physics and philosophy came together in this context has still to be mapped out. There are significant continuities between these developments and current work on not only the metaphysics of quantum theory but also the foundations of space-time theory and quantum gravity. If we can identify and extract these continuities, we might be able to project them forwards and delineate fruitful lines of enquiry in which philosophy and physics can come together and future theories can be shaped in both their physical and philosophical aspects.

Coming back to the foundations of space-time theory, then, the conceptualisation of the underlying manifold in terms of individual points, however these are individuated (either in terms of some form of primitive thisness or via their relations to their fellows), has been seen as underpinning substantivalist views of space-time, which in turn are threatened by the infamous 'Hole Argument'.

Rather than retreat, in the face of this argument, to some form of relationalism, which itself faces acute problems, the suggestion has been made that some form of structuralist approach to space-time might be adopted, according to which, as the great historian and philosopher of physics Howard Stein once suggested some years ago, in a discussion with Adolf Grünbaum, space-time is seen as an aspect of the structure of the world. Again, just as in the quantum case, such an approach has historical precedents, with Eddington, perhaps most famously, adopting such a view of General Relativity (indeed, Stein claims that Newton himself actually held a form of structuralist position).

Constructing a viable theory of quantum gravity then involves tackling the problem of 'background dependence': the fact that, to put it crudely, quantum mechanics is set against a fixed space-time background which raises obvious difficulties in effecting some form of unification with General Relativity. String theory addresses the problem by keeping the fixed background but in the context of extra dimensions which effectively 'delocalize' the interactions between objects away from the space-time points. An alternative set of approaches seeks a background independent theory according to which space-time points are no more than 'nodes' in the metric structure, where the metric is a dynamical entity, represented quantum mechanically via the appropriate operator. The space-time geometry is then itself quantized and, significantly, such approaches seem broadly compatible with structuralist views of both the underlying metaphysics and mathematics (see D. Rickles and S. French, 'Quantum Gravity Meets Structuralism' in The Structuralist foundations of Quantum Gravity). Eddington famously thought that the appropriate mathematics in which such a structuralist vision could be couched was group theory; more recently John Baez has advocated category theory as supplying the relevant formal framework.

Following these early discussions, Rickles, Saatsi and I organized a symposium on these themes at the Biennial Meeting of the Philosophy of Science Association in 2002, to which we invited Baez, the physicist Lee Smolin and John Stachel, a very well known historian and philosopher of space-time physics. On the back of the success of that symposium we put together the book collection, inviting not only other structuralists to contribute, but also critics such as Oliver Pooley, from Oxford. As a representative example of the way a fruitful relationship can be constructed between philosophy and physics, let me mention the essay 'Holism

and Structuralism in Classical and Quantum General Relativity',
by Mauro Dorato, a philosopher, and Massimo Pauri, a physicist.
Here physics and philosophy truly come together through their focus on the meaning of covariance in the context of the Hamiltonian
formulation of General Relativity and their attempt to articu-
late the subsequent metaphysics of space-time in a structuralist
setting. And although I disagree with some of their philosophical conclusions, I thoroughly admire the way in which they fuse
philosophical and physic-al insights into not just the already laid
down core of General Relativity, but the still-under-construction-
foundations of quantum gravity. Here we must resist an updated
version of Eddington's famous admonition that, "It would probably be wiser to nail up over the door of the new quantum theory a
notice, 'Structural alterations in progress—No admittance except
on business,' and particularly to warn the doorkeeper to keep out
prying philosophers" (The Nature of the Physical World, Cambridge University Press.,1928, p. 211).

So, the foundations of quantum gravity is an obviously fertile
field of physics ready for philosophical consideration, but interestingly enough the foundations of statistical mechanics – which
many people may have regarded as 'done and dusted' – is making
a bit of a come-back with some very interesting and through pro-
voking discussions beginning to emerge in the literature. Another
growth area is quantum information theory, which, of course, has
great practical implications as well as offering significant scope for
philosophical analysis. One of my colleagues, Chris Timpson, has
made some important advances in this area, arguing that the term
'information' functions as an abstract, rather than concrete, noun
and hence that quantum information is not part of the material
content of the world (see his forthcoming 'Philosophical Aspects
of Quantum Information Theory'). In particular he claims that
pieces of quantum information are not mysterious or 'unspeakable' in some way, but can be straightforwardly characterized in
terms of particular sequences of Hilbert space states. Such work
is an excellent exemplar of the insight that can be achieved when
a nuanced philosophical analysis is brought to bear on the foundations of physics, even where those foundations are still being
put together, brick by brick.

And, of course, the metaphysics of quantum theory in general
remains an area of fundamental interest for philosophers. It is
interesting to see how the focus has shifted over the last twenty
years or so, from a concern with entanglement and separability

and the like, (back) to the measurement problem and the profound work over the past several years on decoherence and, most recently, new forms of the Everett interpretation. And despite some deep and suggestive work by the likes of Redhead and others in the 1980s, the foundations of quantum field theory deserve further exploration and philosophical articulation.

Personally, I think the most interesting and potentially fruitful project to which philosophers can contribute has to do with the articulation and development of a viable form of realism appropriate for the quantum realm. It is close to scandalous that, on the one hand, certain philosophers seem happy to present forms of scientific realism which at best touch briefly on classical theories or bastardized forms of semi-classical quantum physics and, on the other, physicists present their own ludicrously naïve 'philosophies', often with misplaced pride, couched in a terminology which provokes little more than laughter from the philosophers. It is surely time, past time, for both groups to get over their differences, or professional pride, whatever, and begin to learn from each other so they can all contribute to the greater intellectual good!

7

Nick Huggett

Department of Philosophy
University of Illinois at Chicago, USA

1. Physics and Philosophy

I'm going to answer all the questions at once – in particular those
that ask about the relationship between physics and philosophy,
and how they may influence each other. In fact there are a number of relationships, over time and across different problems; we'll
discuss a few interesting topics to illustrate. (I should say that
I don't intend the examples as a systematic review of the field,
or even of the most important topics.) A note before we start;
by 'philosophy' here I mean – just because of my expertise – philosophy in the so-called 'anglo-american' or 'analytic' tradition
(opposed to various Twentieth Century schools of thought associated with France and Germany in particular). I also generally
exclude 'value theory': ethics and aesthetics for instance (though
again, interesting things could certainly be said.)

First, it's worth saying a few words to explain why physics and
philosophy could have anything in common at all – though quite
possibly the reader of a book like this one will already have a good
idea. There seems to be a gulf between the disciplines because
philosophy often appears to involve argument from premises of
uncertain justification to conclusions of unclear significance, while
physics applies apparently clear empirical methods to well-defined
problems. (It's almost impossible to form consensus in philosophy, while the body of generally agreed on physics grows rapidly.)
In its most extreme, most crude form, this (mis)understanding
of physics and philosophy holds that philosophy is by definition
separated from empirical questions or justification. I've heard
physicists espouse this view: probably they are vaguely recalling
Karl Popper's attempt to demarcate the scientific from the non-
scientific – but Popper never intended to leave all of philosophy on

the non-scientific side of the boundary! (Leaving aside powerful doubts that it can be drawn.)

But while there is plenty of merit in the general characterisations, the picture ignores the common fundamental interests of both physics and philosophy – and, as we shall see at greater length in what follows, why both methodologies are necessary to the common project. For both physics (I'm particularly talking about fundamental physics) and philosophy seek to understand how, at its most basic level, the world is 'put together' – to indulge in hubris, they both want to see the world as the Gods do. (The last question asks why I was drawn to physics and philosophy – that hubris is a large part of the answer.)

Such a project involves, for instance, discovering what kinds of things there are and how they affect one another, but it also involves understanding in some way what it is to be a thing, and what it is to affect something. The first kind of question is more amenable to empirical investigation (understanding how is one of the great achievements of physics), and pretty clearly belongs to physicists; the deepest answers so far are found in particle physics, relativity and cosmology. The second kind of question is of interest to philosophers; how is existence related to spatial location? To continuity over time? How do we refer to things? What does it mean for A to cause B? And so on. However it would be an enormous mistake to think that the philosophical questions are disconnected from physics. On the one hand, many of the answers offered by philosophers make substantial physical assumptions: for instance, do classical concepts of causation apply in a world of quantum 'jumps'? Indeed, it is often the case that philosophical questions can only be well-posed in the context of a physical theory: for instance, is the world deterministic?

On the other hand, it would be unrealistic to pretend that physics proceeds without substantial philosophical commitments – commitments that sometimes have to be abandoned for physics to progress. The most striking example of this point (I bet the most cited example in these essays) is Einstein's reconfiguration of the notions of space and time; Einstein made clear how the concepts with which physicists were operating were not an adequate framework for the new physics that they were discovering, and showed what concepts were. The questions he addressed – what is time? how does it relate to space? what is an instant? – until then seemed philosophical questions. But more than that, Einstein could not address the questions by the usual empirical methods of physics, since they presupposed the old notions of space and time; instead Einstein had to describe new concepts with new logical relations, suitable for the new theories. But that kind of analysis is a perfect example of philosophical method. (Einstein's solo philosophical reconfiguration is of course enormous – most philosophers of physics aim at either more modest goals, or to contribute as part of a community. But the general goals and methods are similar.)

We can put things like this: philosophy – quite generally – is a kind of reflective self-awareness, about self, about action, about values, about reason, and about our deepest presuppositions about the natural (including physical) world. At times this self-awareness is interesting for its own sake, but still must respect our best scientific knowledge; at other times, especially when a major reconfiguring of basic ideas is required for progress, it is crucial to physics; at other times, as in life, self-awareness is a definite hindrance, and physics best makes progress by taking a lot on faith, and pushing on regardless of philosophy. To see what I mean, let's turn to the promised examples.

2. How Philosophy Learns from Physics

There are some philosophical questions that have really obvious connections with physics: for instance, what is space? What is time? What is causation? What is identity? Especially since our understanding of these things has changed (with Copernicus, with Newton, with non-Euclidean geometry, with Einstein, with quantum theories) it is clear that there is no fixed conception of these things which philosophy can take as its subject matter – as our physical knowledge changes, so will philosophical conceptions.

For instance, consider the idea that the universe has not existed forever; in mythological terms, the idea that the universe was created at some time in the past. How should such an idea be understood? Newton believed that time was 'absolute', ticking at a constant rate, regardless of the specific motions of matter. He also believed that there was no first time, just time to the infinite past and future. Thus creation involved bringing into existence of all things at a certain point in infinite time. Newton's contemporary, Gottfried Leibniz instead believed that time was nothing but the sequence of happenings that make up the life of the universe. In that case, if nothing exists, nothing happens, and so there is no time; time starts with creation. One can argue the merits of the views in the context of Newtonian mechanics, but in the general theory of relativity (GTR) – the theory of gravity in terms of curved spacetime – things look very different (a main theme of Hawking's famous book, 'A Brief History of Time').

First, in classical general relativity, we know both experimentally that the universe has its origin in a big bang, and mathematically that that is the norm for similar universes. In such a universe time does not extend infinitely into the past, since in spacetime has its origin at the big bang. But neither is there a first time, instead all and only times after the big bang exist: time is ordered like the collection of instants after noon today – for every instant there is another closer to noon. The discovery in physics of GTR makes a big difference to our understanding of 'creation'.

But second, GTR is only half of our present understanding of the physical world – of the large part, on the scale of bread bins to galaxy clusters (and beyond). The other half is quantum mechanics (QM), which explains small things, from molecules to subatomic particles. At the big bang, the universe, space included, were very small, so both GTR and QM should apply; an account of 'quantum gravity' (QG) is needed. Hawking found in some simple models that if the two theories were both applied, time did not simply disappear into the big bang, as in the classical theory, but instead became 'fuzzy' in a characteristic quantum way. And even, that if one travelled conceptually backwards 'through' the big bang, exchanged roles with space.

Well, that's a fascinating idea – now who's right, Newton or Leibniz? – but one we can't investigate here. The point that I wish to make, is that developments in physics clearly have profound consequences for philosophical questions. (It's true of course that when this happens physicists would like to annex the questions to physics; but it's not clear that can be done completely, and moreover, it is evidence that other apparently hopelessly 'philosophical' questions will become significant to physics.) Physics just is our best understanding of these things, so if philosophers want to be responsible to our best beliefs about the world, then they have to be answerable to physics.

Indeed, such examples suggest an idea that dominates contemporary philosophy of physics: that the method of philosophy of physics is to pose philosophical questions in the language of physical theories and apply mathematics to derive answers. While such a method is absolutely central to work in the foundations of physics – it is a very clear picture of how philosophy can learn from physics – I also want to explain below (◼4) how philosophy of physics can have greater aspirations.

Before we go, another example of physics affecting philosophy; a rather more surprising one, which illustrates, contrary to the beliefs of many philosophers, how deeply the influence of physics can penetrate philosophy.

One of the great engines of philosophy (and again, I mean in the analytic tradition) in the Twentieth Century was the understanding of how language is implicated in philosophical problems. Of course this idea was hardly novel in the history of philosophy, but it became far more powerful because of seismic shifts in the understanding of logic (starting with Frege's work) and linguistics, and because of powerful models of how the new ideas could be applied to philosophical problems: for instance, in the work of Russell, Wittgenstein, Quine, and Kripke. In a sense the paradigm for all this work is Russell's work on definite description; he showed how puzzles concerning the nature of existence could be dissolved by a proper understanding of the definite article, 'the'.

Philosophers of physics work in such a tradition, and apply the same methods as other philosophers, to investigate physical theories. However, there are philosophical cultures in which the assumption is that such methods alone suffice to address philosophical issues. This is probably true in some cases, but it is often a mistake; physics and science more generally also have philosophical import. Let me give an example of what I mean. Followers of Wittgenstein often take as a central insight that any philosophical problem can be solved by proper attention to the way that words are used. (Wittgenstein says things like that but, in a non-expert kind of way, I'm not convinced it's a good description of all his work, which often seems to have insights beyond an analysis of language.) So one might consider our talk about minds and conclude that while it involves temporal locutions ('I'm in a lot of pain right now') it doesn't involve spatial ones (we wouldn't say 'my pain is in this room'); this would be a version of the Cartesian doctrine that the mind is fundamentally non-spatial. Wittgenstein held that view, though what I have sketched here is only the crudest outline of how such reasoning might go – the point is really just to indicate how a certain philosophical approach might lead you to a certain kind of view of the mind.

But in a wonderful, pithy paper, Bob Weingard (1977) pointed out that such an idea is incompatible with what physics teaches us about space and time. For according to relativity theory, space and time are fungible, there are multiple ways of slicing 'space-time' into space and time (naturally associated with the motions of different hypothetical reference bodies). There is no way of abstracting pure time from spacetime, and so no way that minds could be only temporal; the only way that anything can have temporal properties is by having fully spatiotemporal ones. It seems to me that any philosophical approach to the mind is going to have to accommodate this fact somehow, and more generally no philosophical method – no way of applying insights about language for instance – can guarantee that scientific knowledge is irrelevant.

I don't mean that there is no way to defend the non-spatiality of the mind; after all, relativistic effects generally only become observable in bodies moving close to the speed of light, so maybe the mind is part of some 'effective theory' in which time and space are separate. I rather doubt that will work, but my point is that physics cannot simply be ignored here – at least some story of why it is irrelevant is called for, and I don't see how it could possibly be a general story.

Before we move on, let me mention one other important thread in contemporary philosophy of physics, which has become stronger in recent years. That is the study of the history of philosophy of physics. Many philosophers of physics, myself included, spend time not only on foundational problems in contemporary physics, but studying the history of those problems. With our backgrounds in contemporary physics we bring a rather important perspective to the investigation; historical philosophers and physicists are influenced and constrained by many factors, including the physical world itself. But contemporary physics is our best knowledge of the physical world. Moreover, modern physical and mathematical concepts can help clarify the ideas of earlier thinkers – though there is a real danger of imputing anachronistic sophistication to them. We'll talk a bit about Newton below, which will give an idea of the kind of historical work in which philosophers of physics are involved (see also my 'Space from Zeno to Einstein').

3. Physics Without Philosophy

Just as philosophy sometimes progresses without serious reflection on physics – though it can't guarantee isolation – so physics, even fundamental physics, often proceeds without philosophy. (We'll see in the next section that it can't guarantee isolation either.) Particle physics ('quantum field theory', or 'QFT') in the pe-

riod 1940-70 is a very clear example; physicists went from understanding little in detail about processes involving subatomic
particles, to an exquisitely accurate machinery for predicting the
outcomes of their interactions, and the ability to build machines
(well, bombs mostly) harnessing some of the most powerful phenomena in nature. And all this was achieved not only without
conscious philosophical reflection, but in fact with conscious re-
pudiation of philosophy. Steven Weinberg, one of the most visible
and important figures in the latter part of the story entitled one of
the chapters of his book ('Dreams of a Final Theory') 'The Unexpected Uselessness of Philosophy', while the nicest quotation ever
attributed to Richard Feynman (similarly visible and important)
concerning philosophers is: 'Scientists are explorers. Philosophers
are tourists'. Well, they have a point, especially about their fields
and times; but I will explain in the next section why physics has
needed philosophy in the past, and why it is likely to do so again
– indeed, why philosophers can play a role.

This enormous progress was possible for several reasons, prominently among them the following: the basic foundational concepts of QM and relativity were sufficiently well-developed for
the project; physicists with remarkable mathematical insights into
physics were involved; the project could rely on an incredible variety of precise empirical data, both for confirmation and to con-
strain the development of theories. (There are also many interesting sociological and historical factors, including huge military
research and funding, but those are other stories.) Especially because the foundations could be taken for granted, there was little
need for philosophy – indeed, it is probably true to say that too
much consideration of the those foundations would have hindered
progress in QFT (as Weinberg's chapter proposes).

At the same time, issues of interest to Feynman's 'philosophical tourists' were raised by the project (see my 2000 for a sur-
vey). On the one hand, there are some very interesting methodological issues raised: for instance, physicists' far-reaching use of
unproved or even seemingly inconsistent mathematics. For example, there is the need for 'renormalization': the naive algorithm
for producing predictions in QFT produces infinite results, formally because space is continuous. To make sensible predictions,
a new algorithm is introduced which first takes space to be discrete, then adds or subtracts 'counter-terms' to the prediction.
Finally, when one again takes space to be continuous, the coun-
terterms also become infinite, and one is left, formally, with a

finite remainder when infinities are subtracted from infinities. It sounds illegitimate, but the algorithm is well-defined in the sense that the subtraction of infinities is done in a limit. The problem is to understand why renormalization is needed, and why the result is something physically meaningful. I said the questions are interesting to philosophers, but of course these questions were also studied by physicists, especially those concerned with mathematical foundations. (Another issue of this kind concerns the results of Haag to the effect that the standard assumptions of the QFT prediction algorithm entail that particles do not interact. This issue was mostly ignored in the project of QFT, although it did come to be addressed by physicists.)

But during this period, led by more philosophically oriented physicists (who might not have welcomed that description!), philosophers continued to engage with foundational issues in QM especially. The main issues spring from two sources. First there is the issue of measurement; quantum mechanics gives statistical predictions, but because of the way the probabilities enter into the dynamics, quantum probabilities can't (always) be understood as measures of ignorance. The reader may well have heard of Schrödinger's cat: the decay of a radio active atom – a quantum process – determines whether a cat hidden in a box lives or dies. There's nothing very odd about not knowing whether a hidden cat is alive or dead, but that is not the situation. Since life and death are decided by QM, the cat in the box is neither alive, nor dead, nor neither, nor both – it is in a 'superposition'. But the cats we see are either alive or dead, and similarly with anything we experience – it is always some way, not a superposition of different ways. In QM, as usually understood, the story is that it is 'measurement' that explains why that is – but the challenges are to define what a 'measurement' is, and to explain why it does what it does.

Second, there are issues arising from 'entanglement'. Classically, if two things are spatially separated, then they have separate states, and the state of one can affect the state of the other only by the mediation of some kind of force. In QM, it is not always the case that two things can be described by describing them individually – they don't have separate states. Because of this, processes (forces or measurements, for instance) applied to one – at its location that is – will effect the state of both, and so have effects at the location of the other. On the one hand, this fact undermines the idea that we can differentiate things by their locations: do we not just have one thing, bilocated? On the other, the effects, though real, are maddeningly weak – one can't send a message for instance.

In addition, because of the ways that particles entangle, they can also have counter-intuitive collective behaviours, or 'statistics' (this is a topic on which I have worked). For instance, one expects classical particles to behave like dice; if there are ten dice, each of which can take on the numbers 1-6, then there are $6\hat{}10$ possible outcomes to rolling the dice. Or to use an example discussed by Schrödinger, classical particles are like prize medals given to students; if three medals all have different pictures, then are $4\hat{}3$ = 64 ways of giving them to four students – each medal can go to any of the four. But quantum particles do not behave this way. One kind are called 'bosons' after Satyendra Bose. They behave like money on a gift card: if there are three \$10 cards for the four students, then there are only 20 ways of distributing the money. For instance, if two students swapped medals, they would be distributed differently; but if two students swap gift cards it make no difference to the distribution of the money – everyone still has the same amount. The difference here lies in the pictures on the medals; there is nothing like that for the money. Analogously, while even classical particles that are exactly alike – 'identical' in the language of physics – behave as if they have identities that distinguish which has which properties, bosons do not at all; they have no identities that would produce a difference if they swapped their properties.

The issues are surely important, going to the heart of the meaning of QM, but as I said, the answers to the first two in particular were good enough for the purposes of QFT – though they are simply not satisfactory in themselves. (The third issue was more important in some ways in QFT, but there are, I believe, philosophical issues remaining that were not addressed.) What I will suggest at the end is that there are reasons to think that they are important for progress in contemporary physics, and so philosophy is again relevant to physics.

4. What Philosophy Does for Physics

Philosophy interacts with physics most directly in foundational issues – those to do with the most basic concepts of physics, like those just mentioned. Historically, it's always been the case that there are interesting questions to ask about foundations, and so philosophical issues to discuss. But at certain times it simply becomes impossible for physics to progress without serious investigation and revision of those foundations (the same is true in other sciences) – and that work involves philosophical analysis of basic concepts. To be sure, the work has been undertaken by people we think of as physicists – I'm going to discuss Newton and Einstein as important examples – but philosophy is not defined by people in philosophy departments, but as an activity. Anyway, Newton was called a 'natural philosopher', which is not synonymous with physicist – among other things, it does recognise Newton's philosophical work. And Einstein was a 'philosopher-scientist'. Moreover, while these figures worked at a time when great strides could be taken by individuals, today physics is far more of a collective effort. By extension one could think that individual physicist-philosophers will no longer suffice; what is need for today's foundational problems is collaborative work by physicists and philosophers, plural.

In his recent book, Robert DiSalle (2006) studies the history of the concepts of space, time and motion from Newton to GTR, and argues that philosophical analysis of these foundational concepts was crucial for the physics based upon them. I think that he is absolutely right; indeed, the general argument that I want to make in this section, draws on the ideas of that book. He argues that part of Newton's achievement in the 'Mathematical Principles of Natural Philosophy' (the 'Principia') was to explicate clearly (though not quite correctly) the concept of motion in Newtonian mechanics. (That's quite correct, and crucial to understand, though I don't agree with DiSalle that that is all Newton says about spatiotemporal concepts; so my fleshing out of the idea here will be a bit different from DiSalle's.)

We can think about it this way: the goal for Newton's contemporaries was to come up with a theory of mechanics on the model of Euclidean geometry. That is, first one defines various primitive concepts, then one states some axioms, and finally proves the various theorems of mechanics. Descartes' 'Principles of Philosophy' roughly has that structure, though the consequences he claims fall far short of being logical theorems. For one thing, as Newton points out, the way Descartes uses 'motion', is at odds with his definition (actually definitions). For instance, Descartes defines motion as motion from contiguous surroundings, but also says that an object in rotation about an axis tends to recede from that axis. Newton considered a spinning bucket of water; the water is at rest with respect to the bucket, so is at rest by Descartes' definition; but the surface will be concave, demonstrating a tendency to recede from the axis of rotation, so is moving according to Descartes' use of 'rotation'.

In place of Descartes' definition, Newton proposed that 'motion' was change of place with respect to a fixed reference frame, which he called 'absolute space'. That notion is consistent with mechanics – when one says the water rotates, one does not mean in Descartes' sense, with respect to the bucket; one means with respect to absolute space. (I should note for accuracy that Newton models his Principles on Euclid far more closely than Descartes did – and that Newton consciously does not include 'motion' in the definitions.)

Why is this an important achievement? Why was it difficult? Above all, why was it philosophical? It's important, because doing mechanics requires talking about motion, and logically a coherent theory will need a coherent definition. What natural philosophers seemed to find difficult was bracketing various prescientific and philosophical prejudices about space, time and motion.

For instance, Descartes was committed to the idea that all natural phenomena could be reduced to the geometric properties of matter (a crucially important principle, from which modern physics grew); but space itself is something with only geometric properties, so he concluded that space and matter are one. His definitions grew straight from this philosophical conception, and in particular he had no room for Newton's conception of a non-material absolute space. What Newton did was to understand what kind of notion the laws of mechanics required; but that is non-trivial. Instead of coming up with some a priori conception of motion and then trying to formulate laws in terms of it, Newton effectively saw what the laws would have to say and defined motion accordingly. It's one thing to discover some true things in a language, it's another to have also to develop the very language in which to express the things you have discovered. It's the latter, harder task that Newton accomplished. But a big part of what it took was extreme reflexive awareness of role that the concept of motion (and those of space and time) would have to play: and that kind of hard critical analysis of the logic of a concept is philosophy. (Of course, I don't mean to suggest for a moment that the analysis is all Newton needed. Like the creators of QFT described earlier, Newton brought incredible mathematical insights to bear on really useful experimental data, particularly Kepler's

Laws of the planets.)

Einstein's paper 'On the Electrodynamics of Moving Bodies', in which the theory of relativity appears truly for the first time, performs a similar kind of analysis. He showed how a reconceptualisation of our concepts of space and time – especially simultaneity – could make consistent the ideas that (a) light travels at a speed independent of the motion of its source and (b) the relativity of inertial frames means that frames in constant, linear relative motion cannot be experimentally distinguished. (a) and (b) after all imply that light travels at the same speed in all frames! It's not hard to see the similarity with Newton's problem: one needs concepts of space and time in order to state (a) and (b), but the concepts can't be a priori ones, they must be the ones for which (a) and (b) are true. The solution is physics, since (a) and (b) are physical claims, but it is also philosophical, in that it requires an analysis of the role the most fundamental concepts need to play (which is quite different from the naive view). (It's worth pointing out that while the analysis relies on a fairly particular experimental fact – the source independence of the speed of light – it doesn't require any very complicated mathematics.)

That's history, but is there any reason to think that physics does (or will) need a similar philosophical breakthrough? Of course nothingÕs certain, but itÕs reasonable to think that a quantum theory of gravity will require a considerable advance in our understanding of foundational concepts. Let me describe issues in particular – these and other topics are investigated in Callender and Huggett (2001).

On the one hand, there are the unresolved issues discussed above in the foundations of quantum mechanics; Roger Penrose, for instance, has made interesting suggestions about the connection between gravity and the measurement problem. And Carlo Rovelli has also thought carefully about the implications of QM for QG. Now, approaches like theirs are in the minority: the overwhelming effort devoted to the project of QG is given to string theory. But string theory, and thus most research on QG, grew out of the model of QFT; and in particular string theorists do not, in my experience, show a great deal of interest in the foundations of QM.

Here's a reason to think that such an interest would be worthwhile. Unlike Newton's work, unlike relativity, and especially unlike QFT, string theory suffers hugely from a paucity of empirical data. String theorists sometimes respond by pointing out that string theory is the only theory that predicts the existence of gravity from first principles; Newton's theory, for instance, simply postulates the force. But that is to completely understate the role that experimental evidence plays in science. The prediction is nice evidence that string theory is true, but it provides little useful guidance about how to construct a full theory – at present 'string theory' is really only a sketch of a theory. There is nothing like Kepler's Laws, or the light principle, or relativity (or mass scaling, to use an example from QFT), to guide the construction of string theory – as far as I know, there are no big empirical principles, at odds with what is already known, to show the way. (Weingard first made this point in 1988.) Except, there are foundational puzzles about QM; and if one has the idea that a fundamental theory, perhaps string theory, might subsume and account for QM, then those puzzling features start to look like empirical constraints on QG. It is in this spirit that Penrose and Rovelli work.

   The second issue in QG that seems likely to require a philosophical analysis, concerns our conceptions of space and time. It seems pretty clear that a quantum theory of space and time, like QG, will require a radical rethinking of those concepts; perhaps one that can be carried out after the theory is developed, but, going by history, more likely one that has to be carried out as the theory is developed. There's a lot that could be said here, but let me just note that string theorists do seem more interested in these questions. For instance, Ed Witten, the central figure in the field, and Brian Greene (2000), author of 'The Elegant Universe', have written about the implications of 'dualities', symmetries relating models of string theory. One interesting duality is between universes with rolled up dimensions that have very small radii (around the Planck length) and very large radii (like our familiar spatial dimensions, perhaps). They speculate that the duality means that these two models are indistinguishable, and hence – for Witten at least – that perhaps they represent the same situation. But if a tiny space and a large space are the same thing somehow, then isn't that a sense in which space fails to have all the properties it should – a sense in which it 'fades away'? (Philosophers should recognise here something along the lines of Poincaré's conventionalist arguments. Since philosophers know that logical terrain here well, if Witten is right that dualities shed an interesting light on spatiotemporal concepts in string theory, then philosophers should have useful insights to bring to the topic.)

Finally, let me qualify what I've just been saying. I have emphasised here the potential for philosophical involvement in the most speculative, most glamorous, theoretical physics. But I don't want to suggest that is the limit of the involvement, or even the most promising in the next 10 (or 20 years). There are plenty of smaller, more local foundational issues which require philosophical investigation: for instance, in addition to the problems in QM that I have mentioned, philosophers have recently found common ground with physicists investigating quantum information theory.

5. Physics and Philosophy (Again)

What I've attempted to illustrate and argue here is that in addition to the obvious foundational areas in which physics and philosophy intersect, neither can blindly assume that any area is permanently isolated from the other. History confounds that idea, and if one thinks about how progress is sometimes made in the fields (sometimes by assimilating big new ideas from science in the one case, and by conceptual revolution on the other), we can see why. Reflecting on the current state of physics, it's also plausible that physics and philosophy are now in a period of increased mutual relevance.

Citations:

Callender, C. and N. Huggett, (eds.), 2001, Physics Meets Philosophy at the Planck Scale: Contemporary Theories in Quantum Gravity, Cambridge: Cambridge University Press.

DiSalle, R., 2006, Understanding Space-Time: The Philosophical Development of Physics from Newton to Einstein, Cambridge, UK: Cambridge University Press.

Greene, B., 1999, The Elegant Universe: Superstrings, Hidden Dimensions, and the Quest for the Ultimate Theory, New York: W. W. Norton & Company.

Hawking, S. W., 1988, A Brief History of Time: From the Big Bang to Black Holes, New York: Bantam Books.

Huggett, N., 2000, "Philosophical Foundations of Quantum Field Theory", British Journal for the Philosophy of Science, 51: 617-637.

————, 2000, Space from Zeno to Einstein: Classic Readings with a Contemporary Commentary, Cambridge MA: MIT Press.

Schrödinger, E., 1957, "What is an Elementary Particle?", reprinted in Science, Theory and Man, E. Schrödinger (ed.), New York: Dover Publications. 193-223.

Weinberg, S., 1993, Dreams of a Final Theory: The Search for the Fundamental Laws of Nature, Vintage: New York.

Weingard, R., 1988, "A Philosopher Looks at String Theory", reprinted in Callender and Huggett (2001), 138-51.

————, 1977, "Relativity and the Spatiality of Mental Events", Philosophical Studies, 31: 279-284.

Witten, E., 1996, "Reflections on the Fate of Spacetime" reprinted in Callender and Huggett (2001), 125-137.

8

Arthur Fine

Department of Philosophy
University of Washington, USA

Reflections on The Philosophy of Physics

I was drawn into the philosophy of physics by accident. In 1959 I was a graduate student in mathematics working with Karl Menger (a participant in the Vienna Circle and organizer of the concurrent mathematics colloquium in Vienna). I was doing a master's thesis on the structure of the covering theorems of topology and measure theory and I wanted a course on functional analysis, which was not being offered. So I asked Menger if he would supervise a reading course. "What text?," he asked. I suggested we read Paul Halmos' little book on Hilbert space. "No, no, no! We must read von Neumann." He was referring to the forbidding Mathematical Foundations of Quantum Mechanics, which had recently been translated into English (1955). As all students know, that classic text is von Neumann's rigorous antidote to Dirac's impossible "delta functions". It begins by developing the theory of bounded linear operators on Hilbert space and then goes on to build up quantum theory in that mathematical setting. In my studies with Menger we worked through the mathematics but never stopped, reading right on to the infamous no-hidden-variables theorem, and beyond. All along Menger kept raising challenging questions about the interpretation of the physics that von Neumann was developing, the role of his projection postulate, his psycho-physical parallelism, the cogency of his framework for quantum statistics, the no-go theorem, and so on. I recall Menger often exclaiming (with disapproval) that in our day quantum physicists seemed to be the real metaphysicians. I was fascinated and stuck, and I still am! A few years later, now studying philosophy, I had a conversation with Gregor Wenzel, then the Director of the Fermi Institute at the University of Chicago. I wanted to interest Wenzel in my thesis project on quantum measurement theory. Our conversation got off to a bad start when Wenzel warned me that he did not want to hear anything critical about quantum theory, for, as he made clear, the formation of that theory represented his youth! Despite the warning, Wenzel turned out to be a wonderful mentor who helped me understand how physicists think about the quantum theory, which, incidentally, confirmed Menger's metaphysical suspicions.

Philosophy of science in the 1960s was in the doldrums. Postwar logical positivism was going through its most baroque period, producing mostly logical ornamentation in salvage operations directed at every topic from meaning to confirmation. Realism held everyone in a mind-numbing torpor (it still has that effect on some) and the "new" philosophy of science (Norwood R. Hanson, Thomas Kuhn, Stephen Toulmin, et al.) was yet to make a showing. The same was true in philosophy of physics. There were some ongoing disputes in relativity about determinism and the "block" universe, and about the viability of non-standard treatments of simultaneity and clock synchronization. Not the stuff to make your heart flutter. I recall some dreadful discussions about thermodynamics and the arrow of time. In the quantum theory the only extended investigation by a philosopher was Hans Reichenbach's Philosophical Foundations of Quantum Mechanics (1946), which (you guessed it) explored three-valued logic to resolve quantum "causal anomalies". David Bohm's Causality and Chance in Modern Physics (1957) might have been the starting point for an informed critical interest but, like his 1950s revival of de Broglie's pilot wave theory, philosophers paid it almost no attention. Thus even our most advanced thinkers – including Feyerabend and Hanson – were promoting aspects of Copenhagen orthodoxy without realizing that the de Broglie-Bohm theory (which they sometimes mentioned in passing) already provided counterexamples to virtually every argument they were making. The original Bell theorem was published in 1964, but was then little noticed among philosophers (and physicists), similarly for the Bell-Kochen-Specker theorem, which came out in 1966-67. Instead much of the foundations community was caught up in formal analyses of so-called quantum logic, which turned out to be a sideshow. The no-go results began to be understood and explored in the 1970s and 1980s. With them philosophy of physics joined up with new foundational work in physics and began to grow again into a vigorous and relevant discipline. Hilary Putnam's "A Philosopher Looks at Quantum

Mechanics" (1965) is a good example of the state of the art with respect to historical scholarship and conceptual analysis before this revival.

Beginning around 1970 my work related to physics moved in two directions. First I wanted to develop a general framework for the analysis of quantum measurements. This went well beyond my efforts in the dissertation, which ran down a number of misleading claims about measurement in the physics literature by constructing a class of measurement interactions that explicitly contradicted the claims. Rather I hoped to provide a general setting for understanding the roots of the notorious measurement problem. I succeeded in finding a very general framework and with it a correspondingly general no-go result (the "insolubility" theorem). That formed the basis for a later analysis of the role of mixed states in the interaction formalism, and for a viable proposal for how to resolve the measurement problem in terms of what I call "selective" interactions. This work attracted some interest both among physicists and philosophers and has become a fairly standard reference in the measurement literature. My second objective was somewhat less clear. I wanted to understand probability in the quantum domain and I was bothered by the casual treatment of probability in the physics literature, which philosophers mostly mimicked. This led to a study of how random variables are used to represent physical quantities in a probabilistic setting and to my development of an idea of Menger's for the alternative notion of "statistical variables". Unlike random variables, but like the observables of quantum theory, pairs of statistical variables need not have a joint distribution. The failure of joint distributions for non-commuting variables shows up in the "interference" of quantum probabilities, for instance in a typical double slit experiment, and it provides the key to understanding what is going on. Thus confronted with the Bell and Bell-Kochen-Specker (BKS) results I set out to understand them in a way that would display the probability aspects clearly.

This work began with studies on value-definiteness and projectors, where I coined the expression "eigenvalue-eigenstate link" to describe the standard assumption about value-definiteness, an expression that quickly caught on. Departing from the standard interpretation, I did not locate the root difficulty of the BKS theorem in the assumption of non-contextuality, which John Bell had pointed to, but in the algebraic constraints themselves whose extension to non-eigenstates seemed arbitrary. That analysis, however, never caught on, even though I was able to show that imposing the algebraic constraints is equivalent to assuming the framework of random variables with its commitment to the forbidden joint distributions. I had better luck with the Bell theorem itself, where it is now recognized that joint probabilities are the heart of the issue. Theorems of mine, beautifully generalized geometrically by Itamar Pitowsky, show how locality assumptions translate into conditions on joint distributions which are the source of the various Bell-like inequalities. This is the starting point for all the current investigations that use entanglement as a resource for quantum information theory. Here I think that the work of philosophers has had a strong impact on important developments in physics proper. There has been further impact on the design of experiments to test the Bell inequalities. I proposed "prism" and "synchronization" models that accommodate the current generation of experiments without sacrificing locality. Along with the critical analyses of others, these models have prompted experimental technique that explore newly emerging technologies. The object is to devise a class of experiments in which the errors are within bounds (the efficiency loophole) and, simultaneously, superluminal signaling is effectively ruled out (the signal loophole). So far, despite a string of optimistic announcements, no such experiments have been carried out successfully. This raises the fascinating question as to whether what we glimpse here is a new quantum restriction, a kind of uncertainty relation in which efficiency and signaling are bounded reciprocally. In a primitive way, this is what my prism models suggest. It is certainly an area worth further exploration in physics and philosophy.

My investigation of the no-go theorems led me to wonder what the dissidents among the founders of quantum theory had been thinking. That set me off to explore the then virtually untouched Einstein archives and to a decade-long scholarly adventure. Prompted by Kuhn and Toulmin, philosophy of science had begun making an historical turn, flirting with the integration of genuine historical studies into philosophical work. This was mostly done in the form of careful, second-hand case studies. Not since Pierre Duhem and Ernst Mach, however, had there been a tradition of philosophers themselves doing primary historical scholarship. That is what I began to do with respect to Einstein and the quantum theory. John Earman followed with work on Einstein's paths to relativity. These examples, and others, sparked the contemporary investment in history, including the development of a whole subdisciple devoted to the history of philosophy of science. Today, alongside primary work in physics, it is common for philosopher's also to engage in original historical scholarship. Thomas Ryckman's The Reign of Relativity (2005), which brings out the surprising role of Edmund Husserl's phenomenology in the development of relativistic programs, is a splendid example of the historico-analytic genre that has developed, a flowering of what Mach called for in his "epistemological sketches". Slowly our scholarship, along with that of good historians, has begun to filter into the texts and into the background for work in physics proper. Contrary to Kuhn's thesis about the necessity for just-so stories in scientific training, I think that increasingly scientists are open to learning a thick history of their subject. (Unfortunately I can not say the same for their being open to thick sociological descriptions, at least not among most physicists.)

As my own story illustrates, I see the connection between philosophy and science as more or less seamless. In common with other alleged boundaries, there is no hard and fast "line of demarcation" between studying science and doing science. Right now the kind of scientific work that philosophers of physics do is mostly theoretical and often looks a lot like mathematical physics. In part this is a stale leftover from logical positivism, which emphasized formal methods and overemphasized formal rigor. Thus, for example, recent philosophical work on quantum field theory has approached it mainly from the perspective of axiomatic field theory, which is not the framework of most physicists. Most physicists use refinements of the path integral approach which, even in the hands of a master like Edward Witten, is often heuristic and nonrigorous. In my view it would be better for the next generation of philosophers to pursue their philosophical interests with respect to mainstream physics rather than to explore the philosophical implications of frameworks which are not central to how physics is actually practiced. Along these lines, I would also like to see a lot more engagement with experimental practice. In philosophy of psychology, David Buller's Adapting Minds (2005) shows what philosophers can accomplish when they join in the experimental work of their companion discipline. We can do it in philosophy of physics too. For example, Hasok Chang's Inventing Temperature (2004) suggests an agenda related to the measurement of temperature that might offer a useful opening for philosophers to do experimental work in thermodynamics.

There is, to be sure, a conception of philosophy as an armchair, second order discipline squashed between literature, on one side, and empirical science, on the other. But that is a cramped position and I see no reason to suppose one could rest comfortably there – not even in an armchair. So I think we are free on both sides, and certainly free to enliven our understanding of science with our own experimental practice. One of our primary philosophical tasks is to understand how scientific decisions are made. I do not think we can do this well only on the basis of scientific informants and the records they leave behind. We are trying to understand skills and so I think we could also profit by having a stock of communal expertise on which to draw. Thus we should encourage philosophical engagement with science in all its aspects, not just over so-called fundamental theory. My view here differs from another common conception that divides up the intellectual labor temporally. According to this view philosophy engages constructively with science mostly in formative periods (marked by crisis and change), where philosophy can help guide the development of a new paradigm. (On this view, I suppose, while waiting for the next crisis we just can just engage in narcissistic self study.) In my view science is an ever-shifting activity almost always open to change, so I regard good philosophical work as relevant and welcome at all times.

Stemming from the dissident voices of Einstein and Schrödinger, what had been a mostly critical philosophical exercise sorting out "paradoxes" over quantum entanglement is blossoming currently into a rich theoretical and experimental practice in the field of quantum information theory (e.g., quantum computing, key distribution, and teleportation). Because the area is new it offers a golden opportunity for philosophers to join with scientists in theoretical and experimental investigations, which need both critical and constructive attention. For example, using entanglement as a source for the development of technology requires the ability to control decoherence effects, which can quickly destroy the necessary correlations. Up till now philosophers have mainly viewed decoherence with suspicion, but I would guess that the same intelligence that finds critical problems with decoherence schemes can also find ways to utilize them constructively. So the critical decoherence tradition in philosophy may also contribute to a constructive physical practice.

A growth area in physics, of course, is at the interface of the quantum theory and relativity where the community continues its search for the holy grail of a quantum theory of gravity that generalizes the standard model of particle physics and pins down some of its arbitrary features. Myriad problems and puzzles arise here to which philosophy can contribute. Philosophy can also help lend some perspective to string theory and the interesting programs that compete with it. Because string theory is so well entrenched institutionally it would be useful to have some informed, critical reflection on its character and prospects. String theory would be a good laboratory in which to study the epistemological issue of accommodation versus prediction. We might, for example, reflect on the extent to which string theory's indefiniteness and lack of novel predictions represents an anomaly in the history of physics. Studies of this kind might well contribute to the modification and development of several current approaches to quantum gravity.

Another growth opportunity for philosophers is to join in the lively foundational discussions of competing interpretations of quantum theory. The passage of time has loosened the apocalyptic grip of Copenhagen, at least somewhat, and allowed several competing interpretive options to develop and attract the interest of physicists. This includes the de Broglie-Bohm approach, where questions of testability and relativistic compatibility have begun to look interesting, the several collapse theories and the many-worlds approach. The latter seems to have a special lure for theorists in physics while enjoying disfavor among philosophers. It would be good to examine the ideas of these groups together to see whether there is some common ground to winnow out and pursue. There are other proposals that philosophers have not yet studied carefully, including the foundational use of non-commutative geometry and the resources of a purely relational theory. These proposals situate probability in different ways and one of the areas of continued activity in the coming years will be the into the roots of the fundamental probability axiom, the Born rule. Overall, this is a bright time for technically skilled philosophers to pursue a more or less traditional interpretive agenda. My only caution would be to resist the temptation to turn that into an exercise in analytic metaphysics. What can lead one into this trap is pursuit of the Kantian question as to what the world must be like for our scientific theories to be true. Bad question. Bad pursuit. Our rule of thumb as philosophers of science should be to try to connect the issues that we investigate with issues that are relevant to ongoing scientific work. Sometimes, I suppose, that may implicate questions of fundamental ontology. But, I suspect, not often!

There is one more important area for growth that I would like to mention. It is to explore values issues in philosophy of science, including philosophy of physics. As several recent studies have documented, the cold war and the McCarthy period left a legacy of retreat from philosopher's traditional engagement with normative issues, especially in so far as they impact science policy and society more generally. In common with all the sciences, physics involves skilled judgment and hence issues of normativity at every turn. Whether the primary activity is directed at theory or experiment or instrumentation, questions of standards arise and need to be responded to. What experiments to perform, what investigations to pursue, what resources to seek, what data to attend to, what interests to take into account, ... – and how to approach all this. In our daily lives we often find ourselves weighing what to do and how best to act, taking into consideration the many concerned parties. The same is true in the daily life of science. Thus a full range of normative questions are also questions for reflective scientists. and so for us. The standard dodges used to bracket questions of value no longer work. We have come to appreciate, for example, that no firm distinction between the context of discovery and context of justification is viable; the same is true for any general distinction between facts and values. Thus serious philosophy of science can not avoid probing the normative aspects of scientific practice in the same way that we try to understand other aspects of science. Physics, which sometimes like to hold itself out as special, I am afraid is not. We all sail on a normative sea and it this might be a good time to try to make explicit some of our tacit understanding of how we do it – with an eye, of course, to doing it better. John Dewey taught us that in reflective practice itself we can learn how better to pursue it. He emphasized that this learning comes about through the exercise of "intelligence". Surely values are not intelligence-free, nor are they the province of an alien intelligence. So we have work to do.

9

Chuang Liu

Professor

Department of Philosophy

University of Florida, USA

1. What is the relationship between philosophy and physics? What should the relationship be?

I could never forget the pleasant surprise, and a big one, when I first read Aristotle's Physics. I was a third year undergrad in Shanghai who had enough proficiency of English to read the book in English translation, while I had next to zero knowledge of history of philosophy that would have allowed me to anticipate what I might find in it. I was expecting to find primitive generalizations of observations on various different and heterogeneous types of physical phenomena, but instead found what looked like philosophical investigations of fundamental concepts such as space, change, and causes. Subsequent readings of what appeals to a student of physics and mathematics in the history of philosophy convinced me that Russell was quite accurate when he said that not only most branches of modern/contemporary science (natural or social) stemmed from the trunk of metaphysics/philosophy, sprouting away when an empirical methodology is found for investigating a particular set of phenomena and evaluating the findings, but the most fundamental questions remain in the trunk. In this connection I do recommend John Losee's book, which introduces students to philosophy of science by giving a brief account of the history of natural philosophy, and therefore a reading of which goes a long way of putting to rest the doubt about the close historical connection between philosophy and physics.

Of course, one may argue that the trunk-branch metaphor may not be the right metaphor for philosophy and science, that even if most disciplines of modern science did evolve from a common stock known as philosophy, it doesn't follow that they now still hold such a symbiotic relation with philosophy; and the metaphor if taken too far would certainly cease to represent the relation altogether because it is difficult to argue that contemporary disciplines of natural science as branches still draw nutrients from the tree's roots, which are supposed to belong also to philosophy. Or it can be argued that modern science stems from neo-platonism and the tradition of practical arts, such as metallurgy, which ran separately if not parallel to the intellectual history dominated by Aristotelianism. In other words, the claim that modern science has its roots in the 'rational' Aristotelian tradition rather than in the 'mystical' neo-platonist tradition is highly misleading if not false.

Of necessity or not, here is what happened between philosophy and physics in more recent history. There is little doubt that what happened in theoretical physics in late nineteenth and early twentieth century contributed a great deal to the transformation of philosophy. Not only most of the pioneers of contemporary physics reflected philosophically on deep foundational problems in their respective fields, one here thinks of Mach, Boltzmann, Einstein, Planck, Bohr, Heisenberg, among many, many others; and their findings also greatly influenced the formation or transformation of whole philosophical trends, here one thinks of logical positivism in the Vienna Circle, logical empiricism in the Berlin Circle, and perhaps Bertrand Russell himself from the stupor of Hegelian idealism; but also, there is little doubt, the philosophical reflections engaged by the above mentioned pioneers in contemporary physics substantively aided their discovery of new scientific ideas and/or theories. One could, for instance, plausibly argue that the difference between Einstein's work and Lorentz's work regarding special relativity is philosophical and/or foundational rather than scientific. The debate between Mach and Boltzmann and their followers on the issue of atomism is also an excellent example of the symbiotic relation between philosophy and physics. There is a large grain of truth in Kuhn's observation that whenever a crisis occurs in a field of physics, philosophical considerations become prominent among its practitioners trying to find ways to end the crisis. If Russell's metaphor is wrong as suggested above, it would be difficult to explain why such phenomena would occur: how could it be that physicists in crises not only engage philosophical reflections but actually succeed sometimes after such engagements?

Unfortunately, recently history of physics does not seem to show the same tight connection between philosophy and physics as we saw it with Einstein's generation, which prompted some to argue, with some plausibility, that that generation's preoccupation with philosophy has more to do with how physicists in that generation were educated than with any intrinsic relationship between physics and philosophy. While I do agree that social factors have contributed to the change of attitudes towards this relationship, the truth lies less in the claim that Einstein's generation was overly occupied with philosophical reflections but more in that the physicists in (roughly) the second half of twentieth century were too dismissive of them. There are many reasons for this change of attitude among physicists. Success I think might be one of the main factors, and by that I mean both the success in pushing the frontier of physics forward and in making it highly regarded in our society at large. But more importantly and seemingly paradoxically, the establishment and growth of philosophy of science as a profession also contributed to the decline of genuine philosophical interests in the physicist communities. A division of labor usually has the social effect of segregating members of the original undivided group into different sects, the members of each avoid doing and esteeming the things the members of the other do.

Given my understanding of the history of this relationship, it is obvious to me that philosophy and physics are intrinsically joined at the most fundamental level of both, and yet social forces may well prevent practitioners in either philosophy or physics from reaching out to each other. At the moment, the most rational or ideal arrangement in my opinion is to have those physicists who decided to devote their efforts to the research of the most fundamental part of a subfield, whether it be high-energy physics or condensed matter physics or..., well trained both in mathematics and philosophy (training in math is standard today but in philosophy is not), so that they begin their work as much a mathematical physicist as a philosopher. Something of the sort is emerging in recent years, when one witnesses journals such as Foundations of Physics thriving and philosophers of physics regularly publish works in journals such as Physical Reviews, which traditionally belong exclusively to physics.

2. How did philosophers contribute or fail to contribute to the development of physics in the 20th century?

Ernst Mach was both a physicist and a philosopher in the proper sense of both words; and when one reads his "The Science of Mechanics," one has the distinct feeling that he had his philosopher's (and historian's) hat on most of the time when he wrote it. This is at least unmistakably so when one reads the part in which he criticized Newton's notions of absolute space and time; and this was the part that literally influenced Einstein and played a crucial role, according Einstein's own recollection, in his discovery of the theory of relativity, special and general.

Although what I said about this example is largely true, what actually happened in the making of theories of relativity was much more complex. For the fascinating story I recommend Tom Ryckman's book, The Reign of Relativity, (Oxford, 2005).

Other cases in the early part of 20th century physics are less distinctive. Was Bohr also a physicist and a philosopher in the same sense as Mach was? It is not an easy question for me to answer. I can see an argument for an affirmative answer, and yet I am not sure if I can make it if called upon to construct such an argument. But if he was, then so was Einstein and Heisenberg, et al., for there is no sense in which the Einstein-Bohr debates can be construed as debates between a physicist and a philosopher.

We saw in subsequent years a divergence of interests and views between physicists and philosophers of physics/science, largely because of the institutionalization of philosophy of science. While philosophers have mostly concentrated on resolving foundational problems left by the generation of Einstein and Bohr, physicists have moved on, paying little or no attention to what the philosophers had produced or were trying to produce. Some people find this development regrettable because they would like to see the fruit of philosophers' labor directly influencing or giving guidance to physicists' work, but I think it is quite all right. Ideally, it would have been better if physicists and philosophers had been reciprocal in paying attention to each other's works, but in fact, this demand would only make sense even in the ideal circumstances between a small subset of physicists and of philosophers. Only those physicists who work on the most basic parts of their respective fields would directly benefit from interactions with the philosophers who are conversant with foundational problems of physics. In recent years, one does see a trend of physicists and philosophers coming together in tackling the toughest foundational problems in physics, problems such as the quantum measurement problem or the interpretation of quantum theory in general. Still, how much significance such investigations could have on the development of physics in general is not clear and may never become clear because it is never clear how important resolving the foundational problems in a discipline is to its proper development. If Kuhn is right, it has no importance in either the normal-science or the revolutionary periods. A philosophical look at the foundation of a discipline in a crisis mode may be extremely fruitful, but it does not follow that solving any foundational problems will have the same effect. Einstein's philosophical reflections on space and time might indeed have helped him to discover the special theory of relativity, but they did not solve any foundational problems about space and time per se, nor would the solving of any of such problems necessarily help anybody to discover any new and better theories in physics.

So the question is really how much the foundational research in physics contributes or fails to contribute to the development of physics in general, and the answer is clearly "not much" in any direct way. However, unless the term 'foundational' is understood in some twisted or ironic or parochial ways, the contribution of such research to physics must be enormous, even if result-driven research physicists couldn't see it. In other words, I think people should stop wanting to see justification of works in philosophy of physics through some kind of tangible effects of researches in physics at large. Such a justification is never going to be forthcoming because no such justification exists or is needed. The fact that self-reflection seldom helps one to succeed in one's profession (assuming that's true) is no argument for that self-reflection is a waste of time.

Therefore, as I see it, philosophy of physics and physics have different goals as well as use different methods. Working out problems in them satisfies respectively different types of people or different needs in the same person. There lies the separate justifications for pursuing one or the other or both of these two disciplines. Perhaps Machian positivism did in fact 'cause' Einstein to discover his special theory of relativity, but if so, it was a singular causal sequence that conforms to no nomological principles. As we now see, there is no intrinsic connection between positivism and special relativity as far as the content of the two is concerned.

3.   What aspect of current work in physics can benefit the most from collaboration with philosophy?

Any time when you have a shotgun marriage between two theories belonging to two different sub-disciplines, philosophy has a potential role to play. Here I was thinking of Wittgenstein's idea of philosophy as having mainly a therapeutic function. For example, a theory of quantum gravity, where a marriage of quantum field theory and the general relativistic theory of gravitation must take place, and yet it is still a question as to whether quantum theory is compatible with general relativity, or whether it makes sense to quantize gravitational field, which means to quantize spacetime or to see spacetime as yet another quantum field, on a par with e.g. quantum electromagnetic field. As it often happens, when confronted with problems of such magnitude, physicists are not so much short of proposals for a solution as not having good reasons for making a choice among them. The questions may be no more than apparently serious; but to see that it is, philosophical analysis is needed. It is not even clear why a quantum theory of gravity is needed. For this and related problems, see the discussion in Callender and Huggett, eds. Physics Meets Philosophy at the Planck Scale, (Cambridge, 2001).

A case in point, when it is discovered that under the formulation of constrained Hamiltonian systems, a quantum theory of gravity gives us a metaphysical picture of the universe that has no place for change and hence no place for time, a serious interpretational challenge faces physicists and philosophers alike. We don't even know whether such a result comes from the peculiarity of the formulation.

The emergence of quantum physics created a whole host of metaphysical problems that are still turning up new challengers for philosophers of physics. The puzzle of Quantum non-locality associated with Bell's inequalities is one of such problems, and the quantum measurement problem is another. Take the quantum measurement problem for example. Given the canonical formulation of non-relativistic quantum theory which in terms of mathematics uses the linear algebra of Hilbert spaces, one can prove a 'no-go' theorem that effectively prevents any non-practical or non-approximate solution to the problem. In the search for an interpretation of quantum theory that provides a satisfactory solution to this problem, philosophical analysis comes in not so much in coming up with possible alternative interpretations as to provide a framework for evaluating or choosing one interpretation over the other. Because of the enormous empirical success of quantum physics, there is no reason to prefer any interpretations that imply anything that disagrees with the existing empirical results or their extrapolations, any discussions or debates on the interpretations are ipso facto philosophical; even such considerations as invoking aesthetic values, which many physicists treat as something within their professional toolkit, are essentially philosophical.

One of the more recent attempts in resolving the problem might initially be thought of as not involving philosophy at all, but it turns out in close scrutiny that philosophy is deeply involved. The so-called algebraic approach in quantum physics which provides means to describe with complete mathematical rigor macroscopic quantum systems, such as those dealt with in condensed matter physics, as well as quantum fields, provides hope in understanding the quantum measurement process in a new light that is not available in the canonical, single Hilbert-space bound description of quantum systems. The 'no-go' theorem mentioned above becomes inapplicable in this algebraic approach because by casting a quantum system in the abstract algebraic framework, one is not restricted to representing it in a single, separable Hilbert space to begin with; but rather different Hilbert spaces can be used to represent different algebraic states of the system; and one of the consequences of this possibility is a rigorous description of the process through which a coherent state is broken into a thoroughly de-cohered state (or very roughly, a transition from a superposed pure state to a mixed state that is devoid of superposition).

One might think of the above as a solution to the quantum measurement problem without of any use of philosophy; it neither involves a justification for the use of hidden-variables of any kind nor a justificaiton for such ideas as many-world or many-mind. And yet this is not so. The approach 'solves' the problem of quantum measurement only if we takes the bulk-infinity limit and the time-infinity limit for the measurement device and the measurement process respectively. How can such idealizations be justified is precisely the kind of philosophical considerations that only the most pragmatic minded physicists can ignore.

4. What area in contemporary philosophy of physics is most fertile?

Looking at the recent literature in philosophy of physics, one couldn't help getting the following impression. The problems with quantum gravity have been grapping people's attention for sometime. The problem about time – no-change-no-time problem in the constrained Hamiltonian formulation – has generated interesting discussions. Associated problems are, e.g. why general relativity could not be understood and formulated as another Yang-Mill's gauge field. Some efforts have been made by physicists to do just that, but so far no clear success can be seen; and the reason for the lack of that is still not entirely clear.

The quantum measurement problem is still generating new interests. New attentions have recently been paid to Everett's many-world interpretation. Enthusiasm in this direction seems to have replaced the ones for the modal interpretation and the de-coherence approach. Bohmian mechanics, as Bohm's hidden-variable interpretation is now known, still holds some promise. New avenues of approach are starting to attract attentions, e.g. the possibility of solving the problem within the algebraic approach to quantum physics. It hold out the hope that the problem be resolved without in any sense introducing hidden-variables, and thus avoiding the ad hocness associated with such approaches.

The effort to provide a clear understanding of the meaning and significance of the so-called gauge argument – the idea or principle that making a global symmetry of a Hamiltonian a local one introduces an associated field of interaction – is still producing philosophically challenging investigations. This effort is made more important when it is seen as part of the effort to understand the role of symmetry principles in physical sciences. For interesting issues in this connection, I recommend the volume edited by Katherine Brading and Elena Castellani, Symmetries in Physics (Cambridge, 2003). In this connection, a study of the notion of spontaneous symmetry breaking (SSB) is especially important, partly because we now realize that quantum measurements as a type of physical process may well be a species of SSB.

Another large and fertile field of philosophical/foundational research centers on the fault line between the microscopic and the macroscopic. The question of reductional explanation may now look like an out-of-date question, but given the lack of a satisfactory solution to the problem of quantum measurement, our understanding of that relationship is far from conclusive. If the macroscopic, which is largely classical, and the microscopic, which is completely quantum-physical, is still divided by the mysterious processes of quantum measurement, namely, without or before a measurement no classical values of any property of a system obtains, such properties as location, momentum, magnetization, and being alive or dead, the reductional relations physicists routinely assume are in principle utterly incomprehensible. Even if one allows approximate/practical concessions such as using 'semi-classical' or 'semi-quantum' descriptions so that the principal gap is assumed to be absent in the first place, there are still methodological issues, such as what it means and how it is justified to treat a macroscopic system as having an infinite bulk or to take seriously the results obtained 'at the end' of an infinite time.

5. In your opinion, which area of physics holds the most exciting promise in the coming decades?

I am too much of a philosopher, and not enough a physicist, to offer my opinion on such matters. And given what I said above about the relationship between physics and philosophy, I do not think what hold the most exciting promise in physics would/should be something that philosophers of physics necessarily ought to pay attention to, unless one is studying contemporary physics (or science in general) from a sociological or net-work point of view.

6. How were you initially drawn to the field and what are some examples of your work that have influenced the discipline?

I was an undergraduate student in physics in Shanghai when I first discovered the pleasure of reading and thinking about historical and philosophical questions. I remember being led to read Mach's The Science of Mechanics by seeing Einstein's remarks on his experience in reading it; I also remember searching all over Shanghai for a copy of the translation of Thomas Kuhn's book, The Structure of Scientific Revolutions. But despite a taste of philosophy of science that inclined me towards a professional career in it, I was more pushed rather than drawn into the graduate study of philosophy of science. Some of the seniors including myself were allowed to participate in some hands-on research work with our professors, and I was assigned to one who was in the process of writing up a article in high-energy physics. I can no longer recall the exact problem we were trying to solve in that article, but I do remember the task I was assigned to. I was asked to calculate the total mass of a nucleon (containing three quarks), which involved nothing more than a routine calculation of all the elements in a few rather large matrices. I remember spending sleepless nights doing calculations for a whole week, trying to get the calculation done on time. And then more than half a week was spent checking the calculation results of my co-worker on the project. The work was so tedious and the results so messy and even bewildering that it became absolutely clear to me that that was not what I studied physics and mathematics for. I have always been more interested in conceptual issues in physics rather than the clever methods by which certain problems can be solved. I remember once getting a gentle warning from my professor for asking how quantizing a wave-function one more time could turn it - a mere math device for probability calculation - into a functional referring to a genuine quantum field. I was told to pay more attention to how best solve the problems given at the back of our textbook. It is not hard to see how when I read Kuhn I was instantly converted. The admission two years later into the graduate program in the HPS Department at the University of Pittsburgh also had a decisive role in my finally deciding to pursue a career in philosophy of science.

My dissertation, which I've done under the supervision of John Earman and Clark Glymour at Pittsburgh, focused on the history and foundation of relativistic thermodynamics. Even though I let the subject go soon after I embarked on my career, the interest in the relationship between thermodynamic and statistical mechanics stayed with me until today. The way in which I picked my dissertation topic also, unbeknownst to me, became almost the default way I select research subjects in my career. For similar reasons I first chose to study the Aharonov-Bohm effect, and then the problem of phase-transition and critical phenomena, and then the phenomena of spontaneous symmetry breaking in the classical and quantum cases. My analysis of the A-B effect turned out to be just the beginning, scratching only the surface of that theoretical phenomenon. Subsequent works by Healey, Maudlin, and Leeds, among others, show that it not only says something about wave functions, it also says something more profound about properties of a gauge field, such as its realness (or the lack of it) and non-locality. The philosophical interests pertaining to phase-transitions and critical phenomena center on the fact that such thermodynamic properties are only recoverable with rigor through taking the so-called thermodynamic limit, which demands that a finite-volume condensed matter system be considered having an infinite number of degrees of freedom. Getting clear on the philosophical cost of justifying such a limit helps us to understand in part the kind of philosophical risk physicists routinely and nonchalantly undertake for the sake of mathematical rigor. As Craig Callender pointed out in his essay, "Taking Thermodynamics Too Seriously?", there is a trade-off between taking the thermodynamic limit and fully recovering the singular nature of phase transitions and avoiding the limit and treating phase transitions as non-singular; and which option is better justified philosophically is not an easy question to answer. SSB by itself does not seem to engender any philosophical problems (some apparent philosophical puzzlement about such phenomena having now been dispelled), but if it holds the secret and hope of ultimately (re)solving the quantum measurement problem, then its importance for philosophy of physics should be obvious.

I have tried to stay away from the mainstream philosophy of science, and like the subject I picked for my dissertation, the apparent paradox in relativistic thermodynamics, the subjects I have chosen to work on so far in my career have all being specific apparently puzzling if not paradoxical issues in physics that not many philosophers have worked on. I do the same in my work in philosophy of science in general. I am not among the first generation in the study of idealization and approximation in science, but I would like to think of my work there as having made significant progress on previous works and therefore revitalized interests in the area.

10

Tim Maudlin

Professor

Department of Philosophy

Rutgers University, USA

I was recently asked to characterize my philosophical field in under 100 words. The resulting attempt clocked in at only 67:

The relationship between philosophy of physics and physics proper is a delicate matter. Often, they merge imperceptibly into each other. But while the physicist trains like a concert pianist to attack the most complex physical problems using theory, the philosopher is content to learn simply which key corresponds to the written note. He then sets about dismantling the piano, to see how it makes sound at all.

Perhaps I can do no better here than to expand on this observation.

The arena in which philosophy of physics and physics merge imperceptibly is ontology or metaphysics. Ontology is the most generic account of what there is (τ?g$o\nu\tau\alpha$), and since at least some of what there is is physical, at least part of ontology is physical ontology. One might wonder why the study of physical ontology is not simply physics: are not physical theories just hypotheses about what there is, e.g. particles or fields or strings or space-time? But any short acquaintance with the opinions of physicists quickly dispels the notion that when they employ "the same theory", they thereby commit themselves to some particular ontology. To take an obvious example: physicists use the mathematical formalism called "quantum theory", and generally agree about how it is to be employed, but can disagree about even the most basic ontology of the theory. Quantum theory ascribes a "quantum state" or "wavefunction" to a system. But does this mathematical object correspond to anything objective and mind-independent in nature, or does it have irreducible reference to, e.g., the knowledge or information that an observer has about the system? Without an answer to this question an ontological account of the world cannot begin, but physicists– who agree with each other with respect to proper normal scientific practice– may vociferously disagree about the status of the wavefunction. That disagreement seems not to make any difference to normal scientific practice, and so may not be deemed an issue for physics proper. One might usefully think of a philosopher of physics as someone who takes questions like these to be of vital importance.

By this definition the most prominent physicists have generally been philosophers of physics: Newton, Einstein, Bohr, Schrödinger and Bell to name some of the most obvious. Indeed, the most profound philosophers of physics have been professional physicists rather than professional philosophers. But because these sorts of question do not turn on complex mathematical questions but rather on understanding of how the theory represents the world– and what part of the world it represents– the analytical tools employed in the enterprise are those of the philosopher rather than those of the physicist. It just turns out that the best physicists are first-rate philosophers as well.

A short illustration may suffice. The wavefunction of a system appears to represent something about the mind-independent physical state of a system. Certain systems are appropriately represented by certain wavefunctions and others are not, and the division of systems into these classes is a matter of how the systems are. Even those who want to tie wavefunctions to the knowledge of individuals have to concede this, for having knowledge of a system or information about a system requires that the system be some particular way. If a representation is equally appropriate for a system no matter how that system is, then the representation is not (non-trivial) knowledge of it or information about it. Even if different observers with different sorts of information about a system properly ascribe different wavefunctions to it, each wavefunction reflects something about the system, and would not have been an appropriate representation had the system been different.

This naturally raises the question: given that the wavefunction carries some information about a system, does it carry all the information there could be? Can two systems properly ascribed the same wavefunction nonetheless differ physically? If they cannot, then (in an obvious sense) the wavefunction is informationally complete. If they can, it is not informationally complete, and one can rightly ask in what physical respects two systems ascribed the same wavefunction can differ.

It is hard to see how one could approach this question in the abstract: what considerations should suggest that there is or is not anything more to a system than what is represented in the wavefunction? No matter how similar systems appear from the outside, they might nonetheless differ in some respect, although if that respect made no difference to how the system behaves, it would be of no interest to physics. But the question takes on a very sharp form when combined with another question about the dynamics of the wavefunction.

Physicists use particular equations– the Schrödinger equation or the Dirac equation, for example– to calculate the time-evolution of the wavefunction. These equations have the mathematical feature called linearity, that is, if the wavefunction $\Psi(0)$ of a system at time 0 can be expressed as $\alpha\Phi(0) + \beta\Xi(0)$ (with α and β complex numbers, $\Phi(0)$ and $\Xi(0)$ wavefunctions), then $\Psi(t)$ (the state that $\Psi(0)$ will evolve into after time t according to these equations) is $\alpha\Phi(t) + \beta\Xi(t)$. As we say, if the initial wavefunction is a superposition of two other wavefunctions, then the final wavefunction will be a corresponding superposition of the states that the other wavefunctions would have evolved to.

This engenders the so-called measurement problem (although we should note that nothing we have said so far mentions measurement at all). For suppose we create a device so constructed that if we feed in a particle in the state $\Psi1(0)$ the system will evolve with certainty into one in which a cat dies, and if the feed in a particle in state $\Psi2(0)$ the system will evolve into one in which the cat is alive. Such a device is possible in both theory and in practice. Then, by linearity alone, if we feed in a particle in the initial state $\alpha\Psi1(0) + \beta\Psi2(0)$, the wavefunction will evolve into a superposition of a state appropriate to a live and a state appropriate to a dead cat. And if the wavefunction is informationally complete– if every physical fact about the cat is reflected, somehow, in the wavefunction– then the cat itself will apparently not end up either alive (and not dead) or dead (and not alive). This problem arises from nothing more that the assumptions that the wavefunction is informationally complete and that the evolution is always linear. Or, as Bell memorably put it, "Either the wavefunction, as given by the Schrödinger equation, is not everything, or it is not right".

The little argument above displays a conflict between three principles: 1) the wavefunction of an individual system is informationally complete, 2) the wavefunction always evolves linearly and 3) individual cats, at the end of experiments such as that described above, are either alive (and not dead) or dead (and not alive). Since experience seems to verify 3), the live options seem to be denying either 1) or 2). These options have been pursued in various physical theories, and even 3) has been questioned by the "Many Worlds" tradition. But our purpose here is not to further anatomize cogent responses to the measurement problem, it is rather to observe how tremendously simple the problem is. It requires little time to present, and little technical expertise to follow. Nonetheless, many– perhaps even most– physicists either dismiss it as a pseudo-problem or feel convinced that a solution need not deny at least one of 1), 2) and 3).

For example, it is often said that the measurement problem is solved by decoherence, which is a rather complicated property of how the wavefunction evolves (in accord with the linear equations) given certain particular details of the interactions in the system. It is in face of claims like these that the different strengths of the philosopher and the mathematical physicist become manifest. The physicist is likely to go into ever greater mathematical detail in characterizing decoherence or determining when it occurs. The philosopher, on the other hand, will properly insist that so long as the evolution is linear, decoherence cannot possibly solve the problem before us. No matter how much it decoheres, the wavefunction, as given by the Schrödinger equation, is not everything, or it is not right. Decoherence can be an important property in other settings, once the measurement problem has already been solved, but it is of no use at all as a solution to the problem.

Philosophy– or more exactly philosophical method– can assist physics here by laying the logical structure of a problem bare, so one can see what sorts of solution are viable. The key is stripping away the inessential, so one is not distracted by irrelevant technicalities. Philosophy could, in principle, advance the cause of physics in this way, although it is open to doubt whether the work of any professional philosophers has actually had such a salutary effect.

In the other direction, physical theory has certainly provided philosophers with unexpected and unforeseen possibilities and arguments. The combination of space and time into space-time, the geometrization of gravity, Bell's proof of non-locality, and the various interpretations of quantum theory have supplied philosophers with new ontologies, arguments, and explanatory strategies. Just as truth is sometimes stranger than fiction, so the fundamental ontological proposals that have arisen out of scientific practice are even stranger than the untrammeled speculation of philosophers could devise. Acute philosophical analysis serves as a sieve, sorting out those proposals that are cogent from those that are not, the real solutions to problems from the only apparent ones. At its best, philosophical method can help hold physics to standards of clarity and rigor that will improve physical practice itself.

Unfortunately, physics in the 20th century went through several phases that were inimical to good philosophical practice. The theory of Relativity suggested to some (although not to Einstein himself!) that the concepts of physics be given an operational reading, leading to an ultimately sterile attempt to reduce all meaningful talk to a "protocol language". Philosophers were fully complicit in charging into this particular blind alley. The advent and empirical success of quantum theory induced most physicists to turn a deaf ear to Einstein's clear and reasonable complaints and embrace– at least officially– the obscurities of Bohr. And the succeeding generation of physicists, Feynman most famously, adopted outright hostility to philosophy and to the particular field of inquiry that is called "foundations of physics". Feynman said that he never really understood quantum theory, but then provided a shining example of how such absence of understanding need not impede first-rate practice of physics.

It is commonplace nowadays to suggest that the future of philosophy of physics, like the future of physics itself, lies in the theory of quantum gravity, the reconciliation of the Relativistic account of space-time structure with the quantum theory. Such a reconciliation, it is thought, will only come through some radical conceptual innovation which will presumably solve, or at least recast, the measurement problem as we now have it. Perhaps this will come to pass, but I am not sanguine about the prospects. Current approaches to quantum gravity– most notably string theory and loop quantum gravity– do nothing (as far as I can tell) to help with the measurement problem: they simply assume it to be solved, as current practice does. Roger Penrose has proposed an approach that takes the measurement problem seriously, grasping the "it is not right" horn of Bell's dilemma. Penrose tries to tie the non-linear evolution of the wave-function to spatio-temporal structure, and hence to gravity.[1] Other solutions, such as Bohmian

---

[1] See, for example, Penrose's *The Road to Reality* (New York: Knopf,

Mechanics and the GRW theory, provide ways of resolving the measurement problem that have nothing to do with gravity at all. Rather than looking forward to a theory of quantum gravity, these approaches already provide the resources to avoid some conceptual problems that plague more "standard" approaches. (Most notably, the "problem of time" in quantum gravity is supposed to arise because the Hamiltonian in the Wheeler-deWitt equation is identically zero, so a wavefunction governed by such a Hamiltonian would be static. If the wavefunction is informationally complete, then it would follow that the whole physical world is static as well: nothing changes. But in the GRW approach, not all of the evolution is generated by a Hamiltonian, and in the Bohmian approach the "additional variables"– particle positions, for example– can change even though the universal wavefunction is static.[2] So this problem may not arise.)

Insofar as professional philosophy has provided a useful service too physics, it has largely been as a refuge for the study of the foundations of physics. The work of Bell, the significance of Bohm's theory and the GRW theory is more widely appreciated and understood in the community of philosophers of physics than in the community of physicists– we have no "serious work" to distract us!– and we are eager to discuss the sorts conceptual issues that have a hard time being published in physics journals. Some excellent theoretical physicists have even found philosophy departments more congenial environments for their work than physics departments. But a healthy future for both philosophy of physics and for physics itself lies in close attention to foundational problems, not in technical virtuosity. Over sixty years later, I can do no better than defer to the judgment of Einstein, as expressed in a letter to Robert Thornton:

I fully agree with you about the significance and educational value of methodology as well as history and philosophy of science. So many people today – and even professional scientists – seem to me like somebody who has seen thousands of trees but has never seen a forest. A knowledge of the historic and philosophical background gives that kind of independence from prejudices of his generation from which most scientists are suffering. This independence created by philosophical insight is – in my opinion – the mark of distinction between a mere artisan or specialist and

---

2005), chapters 29, 30.

[2]For the Bohmian approach, see Sheldon Goldstein and Stefan Teufel "Quantum Spacetime without Observers" in *Physics Meets Philosophy at the Planck Scale*, Craig Callender and Nick Huggett (eds.) (Cambridge: Cambridge University Press, 2001).

a real seeker after truth.

11

John D. Norton

Center for Philosophy of Science and

Department of History and Philosophy of Science

University of Pittsburgh, USA[1]

1. What is the relationship between philosophy and physics? What should the relationship be?

To someone who does not work in philosophy of physics, it can be hard to distinguish what a theoretical physicist does from what a philosopher of physics does. The differences lie in two areas: their goals and their methods.

The highest goal of theoretical physicists is to find the next theory. That profoundly colors the way they approach foundational ideas. Any idea that aids in finding the next theory is deemed useful. Sometimes the most suggestive ideas are so because they are on the edge of plausibility. However if they show promise of opening new pathways, physicists are quite willing to suspend critical scrutiny. There is no point abandoning a goose about to lay a golden egg because you suspect it may be a turkey! Gold is gold. As a result they may put up with what seems like patent nonsense to a philosopher.

For philosophers of physics, the goal is different. The basic questions remain those asked by philosophers for milennia: What is the nature of space? What is the nature of time? What is the nature of matter? How are things in the world connected? And so on. They seek answers from our best understanding of space, time and matter–modern physics. There is no room for tolerance of fringe thinking for that would compromise the project. They ask: What is our understanding now on the basis of our best science?

---

[1]My thanks to Zvi Biener and Balazs Gyenis for comments.

Philosophy of physics also differs from physics in its method. Philosopher of physics bring the sensibilities of philosophy to physics. To those outside physics, philosophy is synonymous with gazing in wonder at intractable mysteries. To the professional philosopher, the project is just the reverse. It is to take things that are conceptually puzzling and, through rigorous analysis, render them simple and transparent so that the original sense of mystery evaporates. Their method looks to the traditional demands of philosophy that theses must be clearly enunciated and defended by clear and cogent argumentation; and that these demands cannot be compromised. Physicists also value rigor of thought and are used to demanding it throughout their work. However they often relax that same rigor of thought when it comes to the deeper, foundational questions that cannot be settled by some experimental investigation or the proof of a theorem. Precisely because these instruments of decision are unavailable, philosophers of physics redouble the demands for rigor, for now the only real barrier to sloppy thinking is one's own self-discipline.

Einstein once remarked on the ability of our thought and conceptual systems to order experience as a "fact... which leaves us in awe, but which we shall never understand" and "The fact that it [world of sense experience] is comprehensible is a miracle."[2] That is an analysis of last resort for a philosopher, much as a physicist would resist as long as possible the conclusion that some striking physical phenomenon is just plain inexplicable. Take the wonder often expressed over the apparently unreasonable effectiveness of mathematics in physics. Should we not say that it is miraculous that the fundamental truths of a physical theory can be captured in the simplest of mathematical formulae? A philosopher would prefer to suggest, as I and others have,[3] that the perfection of the fit of mathematics to the physical world might be explained more by the creative power of mathematicians, who rework their mathematics retrospectively so that physical laws appear simple in the newer formalisms.

Another example that has recently intrigued me is the notion of being "physical" that is often called up by physicists to rule

---

[2] Albert Einstein, "Physics and Reality," pp. 290-323 in Ideas and Opinions. New York, Bonanza, n.d., on p. 292

[3] John D. Norton, "'Nature in the Realization of the Simplest Conceivable Mathematical Ideas': Einstein and the Canon of Mathematical Simplicity," Studies in the History and Philosophy of Modern Physics, 31 (2000), pp.135-170; on pp. 166-68.

out certain formal possibilities. For physicists, it is a notion with great power that can demand instant and instinctive assent from other physicists. For the philosopher, precisely because of its great power to prohibit without evident grounding, it stands as an authority in urgent need of analysis. As far as I can tell, it is a heterogeneous notion whose real content varies dramatically from context to context. It may merely record a gauge freedom, a case of empirical falsity or a failure to describe a system fully. Each has a different basis and justification. It proves not to be a unified oracular power that transcends ordinary means.[4]

2. How did philosophers contribute or fail to contribute to the development of physics in the 20th century?

It is impossible to give a direct assessment. All physicists work in a larger intellectual environment that has absorbed and continues to absorbs ideas developed by philosophers, just as philosophers in turn draw on new work by physicists. Much of what now counts as truisms for physicists about the relation of physical theory to experience and the "scientific method" were first introduced by philosophers. Einstein remarked in his 1916 obituary of the physicist-philosopher Ernst Mach that "...those who consider themselves to be adversaries of Mach scarcely know how much of Mach's outlook they have, so to speak, absorbed with their mother's milk."

Sometimes the influence can be delineated. Two examples are worth mentioning. In discussing his discovery of special relativity years later, Einstein reported on the importance of earlier philosophical studies: "The type of critical reasoning required for the discovery of this central point [of the illicit character of the absoluteness of simultaneity] was decisively furthered, in my case, especially by the reading of David HumeÕs and Ernst MachÕs philosophical writings." I have urged that the reading of Hume and Mach did not specifically provide ideas on time in this context, but a new view of the nature of concepts.[5] A second well

---

[4] See John D. Norton, "The Dome: An Unexpectedly Simple Failure of Determinism," Prepared for the Symposium "The Vagaries of Determinism and Indeterminism,"

PSA 2006: Philosophy of Science Association Biennial Conference, Vancouver,

November 2006. http://philsci-archive.pitt.edu/archive/00002943/

[5] John D. Norton, "How Hume and Mach Helped Einstein Find Special

known example is the role that philosophers of physics have played in reviving the scrutiny of and proposing solutions to the measurement problem of quantum mechanics. That is work that has been advanced as much by philosophers of physics as by physicists.

Finally I will mention work by John Earman and me, following work by John Stachel, on Einstein's "hole argument." It supplies the clearest statement of how spacetime is treated in general relativity and presents a challenge to any account of quantum gravity that relies on a fixed spacetime background.

3. What aspect of current work in physics can benefit the most from collaboration with philosophy?

This is a question that takes some hubris. For, as a philosopher of physics, I am one step removed from the latest researches in physics and so less likely to know where the advances can be made and what ideas may be useful to those making them. However with that said, I will suggest two areas in which I believe a change of course is called for.

While we have made significant advances in the measurement problem of quantum mechanics, the principal advance consists largely in the sense that the terrain of logical possibility has been thoroughly explored. For, after decades of work by some of the smartest minds, we have yet to achieve a consensus on what the right solution might be. That seems good reason to me to doubt that any solution is the correct one. Recall that the essential problem is to reconcile the linearity of the Schrödinger equation at microscopic scales with the known non-linearity at macroscopic scales. The obvious default is just to assume that the linearity breaks down through some as yet undiscovered physics that becomes active on these larger scales. The alternative that drives the measurement problem literature is the idea that the non-linearity can be evaded somehow by "interpreting" the theory differently. More bluntly, that means that the problem can be made to go away merely by thinking differently about the same equations. Indeed some of the re-thinkings proposed are so extreme as to count as their own reductio ad absurdum. Should we really believe that an unobserved linearity on a macro-scale is so sacred

---

Relativity," in M. Dickson and M. Domski, eds., Synthesis and the Growth of Knowledge: Essays at the Intersection of History, Philosophy, Science, and Mathematics. Open Court, forthcoming.

that it must be saved by imagining that the world of our experience is one of many of a vast multitude of equally real worlds, in which all possible outcomes of measurement are realized? Experience gives us only one world. When we are so desperate as to take such excesses seriously, the time has surely come to revert to the default idea that a non-linearity of who knows what form will intervene on macroscopic scales.

The notion of information has become pervasive in some circles of modern physics. Some of the work attached to it is quite enthralling. Here especially I think of work in quantum computing. It exploits the superposition of quantum theory in a most intriguing way, although it does not illuminate its foundations. Unfortunately much of the information talk elsewhere seems to be confused and, whenever I hear foundational work in which the notion of information figures centrally, I am alerted that extra critical scrutiny will be needed. In my view, the longest lasting excess lies in the literature that proclaims that information theoretic analysis provides a novel exorcism for Maxwell's demon. In joint work with John Earman, I have argued that these exorcisms rely on demonstrations that are circular or groundless[6] and, elsewhere, that work on Landauer's Principle depends on a misapplication of statistical physics.[7] There is a reluctance in the physics community to take these warnings seriously since information theoretic notions seem so fertile. Yet, in my view, decades of theorizing have shown that they are fertile only in producing impressive castles that float in mid-air without sound foundations.

Finally, I have been impressed by the tension between the enthusiastic reports of successes in string theory and the cries of alarm from critics that the theory is no theory at all and has no experimental confirmation. Since the complaints are essentially methodological, I think it would be very useful if philosophers of physics engaged in the problem. However it is hard for a philosopher of physics, who must be one step removed from the physics community, to develop a sufficiently deep understanding of the rapidly changing landscape of string theory.

---

[6] John Earman and John D. Norton, "Exorcist XIV: The Wrath of Maxwell's Demon." Studies in the History and Philosophy of Modern Physics, Part I "From Maxwell to Szilard" 29(1998), pp.435-471; Part II: "From Szilard to Landauer and Beyond," 30(1999), pp.1-40.

[7] John D. Norton, "Eaters of the Lotus: Landauer's Principle and the Return of Maxwell's Demon." Studies in History and Philosophy of Modern Physics, 36 (2005), pp. 375-411.

4. What area in contemporary philosophy of physics is most fertile?

There are many smart people working in all areas and I hold high hopes for them all.

5. In your opinion, which area of physics holds the most exciting promise in the coming decades?

Again, there are many smart people working in all areas. Being a philosopher gives me no special powers of prediction concerning new advances.

6. How were you initially drawn to the field and what are some examples of your work that has influenced the discipline?

My initial interest in philosophy of physics came from a real sense of wonder at the content of modern physical theories. That is a sense I have never lost. I was drawn to graduate work in philosophy of physics rather than in physics since philosophical work let me focus most directly on the foundational issues that fascinated me most.

Over time, my interest in the content of the theories expanded to a fascination with how it was possible, first, for ordinary people to discover amazing results and, second, to have good reasons to believe them. The first fascination led to sustained research into Einstein's discovery of general relativity. That inevitably led to a richer understanding of the second. So I regard those historical investigations as a contribution both to history and also to epistemology. Einstein's discovery of general relativity remains today as one of the signal achievements of modern science; and so the details of how he made his discovery must figure in any epistemology that aspires to do more than tell us what happens when we perceive blue patches.

While readers can find a more complete synopsis of my work on my website (www.pitt.edu/~jdnorton), the most lasting contribution of my work was the analysis of Einstein's Zurich notebook. The notebook contains the scratch pad calculations Einstein made during a decisive phase of his work on general relativity and provides a quite fine-grained reconstruction of the course of his discovery.[8] The analysis of Einstein's pathway to general relativity led to another contribution that I believe has proven useful: a sustained examination of Einstein's ideas on general covariance and the protracted debates over them that followed.[9]

Finally, I believe that the analysis John Earman and I gave of Einstein's hole argument has provided a template for later analysis of gauge freedoms and the criteria used to decide when a formal equivalence betokens physical equivalence. The two conditions we described remains those in use today: verification and determinism.[10]

---

[8] See John D. Norton, "How Einstein Found His Field Equations: 1912-1915," Historical Studies in the Physical Sciences, 14 (1984), pp. 253-315; reprinted in D. Howard and J. Stachel (eds.), Einstein and the History of General Relativity: Einstein Studies Vol. I, Boston: Birkhauser, pp101-159; and more recently, Michel Janssen, John D. Norton, Juergen Renn, Tilman Sauer, Michel Janssen, John Stachel, Commentary in The Genesis of General Relativity: Documents and Interpretation. Vol. 1. General Relativity in the Making: Einstein's Zurich Notebook. Dordrecht: Kluwer.

[9] John D. Norton, "General Covariance and the Foundations of General Relativity: Eight Decades of Dispute," Reports on Progress in Physics, 56 , pp.791-858.

[10] John Earman and John D. Norton, "What Price Spacetime Substantivalism? The Hole Story," British Journal for the Philosophy of Science, 38 , pp. 515-25.

12

Laurent Nottale

CNRS, LUTH

Observatoire de Paris-Meudon, France

In your opinion, which area of physics holds the most exciting promise in the coming decades?

What is the relationship between philosophy and physics ?

In a letter to Pauli of 1948 [1], Einstein wrote (author's translation), about the question of the completeness of quantum mechanics in connection with the principle of relativity :

" [...] this complete description could not be content with the fundamental concepts used in point mechanics. I have told you more than once that I am an inveterate supporter, not of differential equations, but quite of the principle of general relativity, whose heuristic force is indispensable to us. However, despite much research, I have not succeeded in satisfying the principle of general relativity in another way than thanks to differential equations; maybe someone will find out another possibility, provided he searches with enough perseverance. "

In my opinion, the " heuristic force of the principle of relativity ", pointed out by Einstein in this letter, remains the most exciting promise in the coming decades, and actually at any time and for any science. Indeed, the principle of relativity, in its general acceptance, goes beyond its application to given particular "objects", properties or theories [2], and can be considered to be the "mother" of all other principles and laws. It is basically a philosophical principle, which can be subsequently translated into a physical principle (covariance and /or equivalence principle), then into mathematical tools and methods (covariant derivatives, geodesic equations). The physical theories so constructed are meta-theories rather than particular ones, as can be seen from the fact that, at each new step of the history of relativity, it is the whole of physics which should be rendered "relativistic". As such, the principle of relativity is at the heart of the relations between philosophy and physics, which can be traced, as concerns modern physics, back to Galileo's metaphor about the "book of nature written in mathematical language" [3] and Newton's "Mathematical principles of Natural Philosophy", in which questions that may have been thought to be of philosophical nature can be turned into physics thanks to the mathematical description tool. Under such a view, a metaphysical question of a given epoch may later become physical, then may find a physical answer.

A fundamental example of this close relation between philosophy and physics is given by the way the principle of relativity allows one to reformulate Leibniz's questions. Leibniz, in his "Principles of Nature and Grace", asked two fundamental questions:

"Why is there something rather than nothing ? Because the nothing is simpler and easier than something.

Moreover, assuming that things should exist, one needs to understand why they must exist in this way, and not in another way."

It is generally considered that the first question does not come under the realm of physics, nor of any science, but of metaphysics, while the primary goal of physics would just be to answer to the second (the "how", but not the "why").

However, several decades before Leibniz, Galileo unveiled the nature of motion and set in the same bound the great program of natural philosophy (the "book" of nature) which was to become, under the impulse of Descartes and Newton, what we call today physics. In order to explain his discovery of the relativity of inertial motion, Galileo wrote in the "Dialog about the two great systems of the world" [4]

"Motion is motion and acts as a motion as long as it is in relation with things which are devoid of it. For all things that take part equally in it, it does not act, it is as if it were not. [...] Motion is as nothing."

Then Galileo erected this finding to a principle:

"Let us therefore set as a principle [...] that, whatever be the motion that one attributes to the Earth, it is necessary that, for us who [...] partake of it, it remains perfectly imperceptible and as not being."

What does become, in these conditions, the first of Leibniz's questions, applied to this "thing" that is motion. Is there really something rather than nothing ? Such a primordial question, seeming to come under the only philosophical thought, is invalidated in its very statement by the discovery which has founded modern physics. Motion, Galileo finds, is simultaneously "nothing" and "something", it is "something" in a relative way, but there exists no absolute motion. It is therefore the very mode of being that relativity questions, while Leibniz takes for granted that a thing is either existent or non-existent. Now, we now know from relativity theories that motion (velocity, acceleration), gravitation, energy, momentum, and many other physical quantities, some of which are even invariants, are existing only relatively to a given reference system, while they are, at the same time and without changing them, non-existing in a particular reference frame which is but the proper one. In the theory of scale relativity, it is suggested that other fundamental quantities such as mass, spin and charges also come under relativity, i.e., they have no intrinsic or absolute existence but depend on the state (of scale) of the reference system and vanish in their proper frame.

It is remarkable that the Galileo statement has been anticipated by the principle of relativity-emptiness of Buddhist philosophy, according to which the various phenomena are empty of proper (absolute, intrinsic) existence and appear only in a relative and interdependent way. For example, one finds in Nagarjuna's 2000 years old writings [5] an analysis of the relativity of motion (among other phenomena) which is almost word by word identical to Galileo's analysis.

Einstein's first insight about the relativity of gravitation, which he has qualified " the most fortunate of all his life " is of a similar nature. Namely, he realized in 1907 that if a man moves in free fall in a gravitational field, he no longer feels his own weight (see, e.g., [6]). Einstein has then translated this first statement into the principle of equivalence, according to which a gravitational field and a uniform acceleration field are locally equivalent, so that a gravitational field can be locally canceled (or reversely, an apparent gravity field can be created) by the choice of an accelerating reference system.

In both cases, the key point raised by Galileo and Einstein is, both for motion and gravitation, their absence of absolute existence. The velocity of a body, its acceleration, and the gravitational force felt by a body are not intrinsic or local physical properties of the body, but relative properties that depend on the reference system. They are not individual properties of objects, but inter-properties between objects, which have therefore the status of relations. By changing the reference system, the property changes (this is relativity), and in the limit there is one reference system, namely, the proper reference system, in which the property vanishes (this is " emptiness ").

Another highly relevant general remark about the nature of the principle of relativity concerns its connection with space-time theories. It is remarkable that all successive theories of relativity (Galileo's relativity, Einstein-Poincaré special relativity,Einstein general relativity, and now scale relativity) are naturally achieved in terms of progress in the description of the geometry of space-time (respectively Euclidean space, Minkowskian and Riemannian space-times, then non-differentiable fractal space-times). This is easily understandable if one realizes that space-time is defined as an inter-relational level of description between objects, as it is manifest in the metric of general relativity which is a generalization of Pythagoras' relation. The account of more and more general transformations between coordinate systems therefore naturally leads to more and more general geometries of space-time.

This general view of the principle of relativity and of its link to the evolution of space-time geometric theories reveals to be extremely powerful. Up to now, relative properties of physical objects such as position, orientation, velocity and acceleration are incorporated in relativity theories. The inclusion of gravitation in Einstein's general relativity (of motion) has marked another step for physics, since gravitation was considered since Newton as a universal field, having an intrinsic existence.

But there are nowadays yet many properties of physical objects (masses, spin, charges of elementary quantum particles), fields (electromagnetic, weak and strong fields) and apparently universal phenomena (quantum laws) which may seem not to come under the relativity principle (which is different from the question of their consistency with special and general motion relativity), in the above meaning of founding them from the principle of relativity-emptiness, and of being able to define the proper reference system in which they vanish.

Indeed, while the principle of relativity (of motion) underlies the foundation of most of classical physics, quantum mechanics, though it is harmoniously combined with special relativity in the framework of relativistic quantum mechanics and quantum field theories, seems, up to now, to be founded on different grounds. Actually, its present foundation is mainly axiomatic, i.e., it is based on postulates and rules which are not derived from any underlying more fundamental principle.

The theory of scale relativity suggests an original solution to this fundamental problem. Namely, in its framework quantum mechanics may indeed be founded on the principle of relativity itself, provided this principle (previously applied to position, orientation and motion) be extended to scales. One generalizes the definition of reference systems by including variables characterizing their scale, then one generalizes the possible transformations of these reference systems by adding, to the relative transformations already accounted for (translation, velocity and acceleration of the origin, rotation of the axes), the transformations of these variables, namely, their relative dilations and contractions. In the framework of such a newly generalized relativity theory, the laws of physics may be given a general form that transcends and includes both the classical and the quantum laws, allowing in particular to study in a renewed way the poorly understood nature of the classical to quantum transition.

A related important concern is the question of the geometry of space-time at all scales. In analogy with Einstein's construction of general relativity of motion, which is based on the generalization of flat space-times to curved Riemannian geometry, it is suggested, in the new framework of scale relativity, that a new generalization of the description of space-time is now needed, toward a nondifferentiable and fractal geometry. New mathematical and physical tools are therefore developed in order to implement such a generalized description, which goes far beyond the standard view of differentiable manifolds.

Let us indeed briefly recall the fundamental principles that underlie, since the work of Poincaré [7] and Einstein [8,9], the foundations of relativity theories, and which can now be applied, in addition to the relativity of motion considered by their works, to the new relativity of scales. We shall express them here under a general form that transcends particular theories of relativity, namely, they can be applied to the transformation of any variable characterizing the state of reference systems (origin, orientation, motion, scale, etc...).

The basic principle is the principle of relativity, which requires that the laws of physics should be of such a nature that they apply for any state of the reference system. In other words, it means that physical quantities are not defined in an absolute way, but are instead relative to a reference system. This principle (which is still of philosophical nature at this stage) is subsequently implemented in physics by three related and interconnected principles (covariance, equivalence and geodesic principles), which are equipped with their corresponding mathematical tools (covariant derivatives, tensor and operator calculus).

(1) Principle of covariance

It requires that the equations of physics keep their form under changes of the state of the reference systems. As remarked by Weinberg [10], it should not be interpreted in terms of simply providing the most general (arbitrarily complicated) form to the equations, which would be meaningless. It rather means that, knowing that the fundamental equations of physics have a simple form in some particular coordinate systems, they will keep this simple form when considering more general coordinate systems. With this meaning in mind, two levels of covariance can be defined:

(i) Strong covariance, according to which one recovers the simplest possible form of the equations, which is the Galilean form they have in the vacuum devoid of any force. For example, under this principle, the equations of motion in general relativity take the free inertial form (which states that the acceleration vanishes) in terms of Einstein's covariant derivative, so they come under strong covariance. One of the main tools for implementing strong covariance is the tensor calculus, which is a natural generalization of vectors to several indices, and which allows to compactify the writing of the equations [11,12], namely, tensors have a particularly simple way to transform under changes of coordinate systems.

(ii)Weak covariance, according to which the equations keep the same simple form under any coordinate transformation, but not as simple as the free Galilean-like equation. A large part of the general relativity theory is only weakly covariant in this context. For example, Einstein's field equations have a source term, the stress-energy tensor of matter, while Einstein's initial hope was to construct a purely geometric theory in which the sources themselves would be of geometric origin. Another case of weak covariance in general relativity concerns the gravitational fields (the Christoffel symbols), which are not tensors, but transform themselves under changes of coordinates in a more complicated, non-linear way.

(2) Principle of equivalence

It is a more specific statement of the principle of relativity, when it is applied to a given physical domain. In general relativity, it states that a gravitational field is locally equivalent to an acceleration field, i.e., it expresses that the very existence of gravitation is relative to the choice of the reference systems, and it specifies the nature of the coordinate systems in which gravitation locally disappears. Therefore, in such an accelerating coordinate system, the field of gravitation vanishes, so that one recovers the strongly covariant form of free motion devoid of any force, i.e., expressed once again by a vanishing acceleration.

In scale relativity [13], one may make a similar proposal and set down generalized equivalence principles according to which the quantum behaviour is locally equivalent to a fractal and nondifferentiable motion [14], while the gauge fields are locally equivalent to expansions or contractions of the internal resolution variables needed to describe a nondifferentiable manifold [15].

(3) Geodesic principle

It states that the free trajectories are the geodesics of space-time. It plays a very important role in a geometric theory of relativity, since it means that the fundamental equation of dynamics is completely determined by the geometry of space-time, and therefore has not to be set down as an independent equation. Moreover, in such a framework the action can be identified (modulo a constant) with the fundamental metric invariant which is the proper time itself, so that the action principle becomes nothing else than the geodesic principle. As a consequence, its meaning becomes very clear and simple: namely, the physical trajectories are those which minimize the proper time, i.e., it becomes a general Fermat-Maupertuis principle. Moreover, contrarily to what happens in other fundamental physical theories, which need to specify both the field equations and the motion (i.e., dynamics) equations (such as in Newton's and Maxwell's theories), once the geometry is known (i.e., the field) the trajectories are also known since the geodesics are completely defined by the geometry.

One may even go one step further in the scale relativity framework. While in general relativity one still considers that the geodesics describe the trajectories followed by 'particles' (of matter or radiation), one may identify in scale relativity the particles with the geodesics themselves (see e.g. [16]. They are therefore no longer viewed as trajectories of 'something', but only as purely geometrical paths (in continuity with Feynman's path integral view [17]), from which the various properties of the wave-particle emerge.

There is another form of the geodesic principle which is a direct consequence of the relativity principle, under its 'emptiness' formulation. It amounts to generalize the initial Einstein statement that led him to the principle of equivalence for gravitation, according to which, "for someone who is in free fall in a gravitational field, the acceleration of gravity vanishes". In other words, in a reference system which is driven with the trajectory, which is nothing but the geodesic reference system, the various forces disappear and the local expression of the motion on the geodesical line takes the Galilean form of rectilinear inertial uniform motion. One recovers the result obtained from the strong covariance and equivalence principles. But here, the meaning is that of a geodesic equation, since the geodesics of an Euclidean space are indeed straight lines travelled through with uniform velocity.

Covariant derivative

The concept of covariant derivative is the main tool designed by Einstein and generalized in scale relativity in order to implement these principles. This tool includes in an internal way the effects of geometry through a new definition of the derivative, contrary to the standard field approach whose effects are considered to be externally applied to the system.

In general relativity, it amounts to substract the geometric effects to the total increase of a vector, leaving only the inertial part (see, e.g., [18]). The same is true in the geometric theory of gauge fields one can build in the scale relativity framework [15,19]. As regards the "quantum-covariant" derivative introduced to account for non-differentiable geometry in the scale relativity theory[10,11], it is of a different nature: the new terms added in its expression are actually a generalization of the expression for total derivatives. Anyway, in all cases these tools merits the name of "covariant derivative", under the meaning of being able to implement the strong covariance principle, i.e., of writing the equations of physics (of gravitation in Einstein's general motion relativity, of quantum mechanics and gauge fields in scale relativity) under their Galilean free form in vacuum.

Methods of scale relativity theory

Since the basic method of scale relativity amounts to explicitly introduce scale variables ("resolution") in the physical description, its construction can be decomposed into three steps.

(i) Find the laws of scale at a given point and instant. These laws are obtained as solutions of differential equations acting in scale space (i.e., solutions that describe the effect on physical quantities of an infinitesimal zoom), constrained by the principle of scale relativity.

(ii) For each of these underlying scale laws, find the laws of motion in standard space(-time), i.e., the fundamental equations of dynamics. They are written in terms of a geodesic equation using a covariant derivative tool that includes the effects of nondifferentiability and fractality in the differentiation process itself. The laws of motion constructed in this way acquire a quantum-type form [13,14,17].

The theory of scale relativity therefore follows a line of thought similar to general motion relativity, and constructs new covariant tools enabling the description of the geodesics of a nondifferentiable and fractal space-time geometry founded on the principle of the relativity of scales.

(iii) Find the laws of coupling of scale and motion. In this case the scale variables become themselves functions of coordinates. These resulting scale fields, which are manifestations of the fractal geometry, can actually be identified to gauge fields of the Abelian and non-Abelian types [15,19].

How were you initially drawn to the field and what are some examples of your work that has influenced the discipline ?

I made in the 70's the remark that the failure of a large number of attempts to understand the quantum behavior in terms of standard differentiable geometry indicated that a possible 'quantum geometry' should be of a completely new nature. Moreover, following the lessons of Einstein's construction of a geometric theory of gravitation, it was clear that any geometric property to be attributed to space-time itself, and not only to particular objects or systems, was necessarily universal.

Fortunately, the founders of quantum theory had already brought to light a universal and fundamental behavior of the quantum realm, in opposition to the classical world, namely, the explicit dependence of the measurement results on the apparatus resolution, as described by the Heisenberg uncertainty relations.

This motivated me to ask the following two questions: Was it possible to describe intrinsically a space-time whose geometry would be explicitly dependent on the scale of observation ? Could such a geometry be able to give rise to the quantum behavior of the objects embedded into it and participating to it ?

Now the concept of a scale-dependent geometry (at the level of objects and media) had already been introduced and developed by Benoit Mandelbrot, who coined the word 'fractal' to describe it [20,21]. The new program amounted to use fractal geometry, not only for describing 'objects' (that remain embedded in an

Euclidean space), but also and mainly for describing in an intrinsic way the geometry of space-time itself.

Basing myself on these reflections, I was led in the years 1979-1980 to suggest that the quantum properties could be the result of a new manifestation of the principle of relativity, generalized to a scale-dependent, i.e., fractal, geometry of space-time [22]. Another complementary line of thought leading to the same suggestion comes, not from relativity and space-time theories, but from an analysis of the quantum mechanical behavior itself. Indeed, it has been discovered by Feynman [17] that the typical quantum mechanical paths (i.e., those that contribute in a main way to the path integral) are nondifferentiable and fractal. Namely, Feynman has proved that, although a mean velocity can be defined for them, no mean mean-square velocity exists at any point, since it diverges in a way which has been subsequently identified with a fractal behavior of fractal dimension 2 (the same as that typical of Brownian motion). Based on these premises, the reverse proposal, according to which the laws of quantum physics find their very origin in the fractal geometry of space-time has been suggested and developed along the scale relativity approach [22,13-16] and Ord's approach [23] that extends the Feynman chessboard model and works in terms of probabilistic models, in the framework of the statistical mechanics of binary random walks.

There is however an important difference between these two approaches, in particular from the philosophical point of view. The Ord approach to quantum mechanics is based on the hypothesis that space-time is fractal and does not come under relativity theories. On the contrary, in analogy with Einstein's method, in which the assumption of space-time flatness is given up and the concept of space-time generalized to curved space-times (which do contain, as a particular case, flat Minkowskian space-times), one also proceeds in scale relativity theory, not by adding hypotheses, but by decreasing the number of fundamental postulates. Namely, while most of physics assumes space-time continuity and (two times) differentiability (except possibly at some local singularities), the scale relativity theory is based on the giving up of the differentiability postulate (while keeping continuity). In such a generalized framework, differentiable space-times are recovered as particular cases, and the fractality of space-time is no longer an hypothesis, but can be proved (this is a mathematical theorem) as a spontaneous manifestation of continuity and nondifferentiability (see [24,25] for studies of the relations of scale relativity with philosophy, in particular in connection with Bachelard's inductive value of relativity, Simondon's realism of relations and the structuring power of the giving-up of postulates).

References

[1] Einstein A., 1948, Letter to Pauli, in Albert Einstein, Oeuvres choisies, I, Quanta, Seuil/CNRS, p. 249.

[2] Levy-Leblond J.M., 1976, Am. J. Phys. 44, 271.

[3] see Alunni Ch., Codex Naturae et Libro della Natura chez Campanella et Galilée, Annali della Scuola Normale Superiore di Pisa, Série III, Vol. XII,1 (1982).

[4] Galileo Galilei, 1630, Dialogo sopra i massimi sistemi, Torino, 1975. (Dialogue sur les deux grands systèmes du monde, Ed. du Seuil).

[5] Nagarjuna, in Kalupahana D.J. (trad.), 1986, " Nagarjuna: The Philosophy of the Middle Way " State University Press of New York.

[6] Pais A., 1982, Subtle Is the Lord: the Science and Life of Albert Einstein, Oxford University Press, New-York.

[7] Poincaré H., 1905, C. R. Acad. Sci. Paris 140, 1504.

[8] Einstein A., 1905, Annalen der Physik 17, 891.

[9] Einstein A., 1916, Annalen der Physik 49, 769, translated in The Principle of Relativity (Dover, 1923, 1952), p. 111.

[10] Weinberg S., 1972, Gravitation and cosmology, John Wiley \& Sons, New York.

[11] Alunni Ch., 2001, Revue de Synthèse 122, 147.

[12] Alunni C. & Nottale L., 2007, " Pour une diagrammatique catégoriale. Mélanges pour Gilles Châtelet ", Colloque en l'honneur de Gilles Châtelet, Revue de Synthèse, Springer, sous presse

[13] Nottale, L., 1993, Fractal Space-Time and Microphysics: Towards a Theory of Scale Relativity, World Scientific (347 pp.)

[14] Nottale L. & Célérier M.N., 2007, J. Phys. A: Math. Theor. 40, 14471

[15] Nottale L., Célérier M.N., Lehner T., 2006, J. Math. Phys. 47, 032303

[16] Nottale L., 2008, " Fractals in the Quantum Theory of Space-time ", in Encyclopedia of Complexity and Systems Science, Springer, in press

[17] Feynman R.P. & Hibbs A.R., 1965, Quantum Mechanics and Path Integrals, (MacGraw-Hill, New York).

[18] Landau L. & Lifchitz E., 1970, Field Theory, Mir, Moscow.

[19] Nottale L., 2007, " Foundation of gauge field theories on the principle of scale relativity " , in Proceedings of International Colloquium " Albert Einstein and Hermann Weyl: 50th anniversary of their death, open epistemologic questions " (Lecce, Italy, 2005), in press,

[20] Mandelbrot B., 1975, Les Objets Fractals, Flammarion, Paris.

[21] Mandelbrot B., 1982, The Fractal Geometry of Nature, Freeman, San Francisco.

[22] Nottale L., 1981, " Fractals and Nonstandard Analysis: a model for the microstructure of space ", preprint, published in Nottale L. & Schneider J., 1984, J. Math. Phys. 25, 1296 and Nottale L., 1989, Int. J. Mod. Phys. A 4, 5047.

[23] Ord G.N., 1983, J. Phys. A: Math. Gen. 16, 1869.

[24] Nottale L., Barthélémy J.H., Bontems V., Walter C., Rosental P.A., de Swaan A, Alunni C. and Brian E, 2001, Revue de Synthèse T. 122, 4th S., Number 1, Albin Michel, Paris.

13

Roland Omnès

Laboratoire de Physique Théorique
Université de Paris-Sud, Centre D'Orsay, France

---

1. What is the relationship between philosophy and physics?

This is typically a philosopher's question. Leaving aside a number of standard considerations that can be found in textbooks, I believe that the essential relationship between philosophy and physics belongs to their common relation with the laws of nature. There are of course many other aspects to their relationship, but many are due to the coexistence of different university departments with different curricula. They cannot be considered as negligible, but I shall try to restrict my answers in the present case to "the thing itself", as far as possible.

1.5 What should their relation be?

This is a difficult question, particularly when stated at the level I propose. Very grossly, one might say that a person behaves as a physicist (or a mathematician) when trying to extend our knowledge of the laws of nature in more depth, precision or coverage. He or she is behaving as a philosopher when considering this knowledge from one or the other of two very different standpoints. There is the standpoint of critique, which is always very important though, if I may say, rather standard. Another standpoint has always been the privilege of great thinkers, notwithstanding their specialty. One may call it a vision, or the creation of a conceptual framework, which would consist first, in the present case, in assessing the meaning of the word "existence" when speaking of the existence of the laws of nature (after a due discussion, of course, of induction, falsification, and a revision of Kant's restrictions on the limits of metaphysics when considering this existence;

and so on). I am really impressed by the shyness of philosophy regarding this endeavor, which is so obviously lacking.

2. How did philosophers contribute or fail to contribute to the development of physics in the 20th century?

I am unfortunately unable to quote any essential contribution from philosophers to the development of physics, although there have been of course many subtle or clever books. This evaluation should be considered as a personal opinion and it means only that I had no occasion to learn something useful for my own understanding of nature from such books, although they were sometimes an enjoyable reading. Popper is for instance the man one thinks of in the first place, but his idea of falsification was an extension of a procedure he had learned from physics and no physicist has yet found any use for his overvalued considerations about probabilities. He contributed undoubtedly to philosophy and human sciences, but not to the development of physics, as far as I can appreciate.

One cannot say however that philosophers failed to contribute to the development of physics, because they were not expected to do so, except for the vision I mentioned previously. There were great thinkers at the beginning of 20th century (Russell, Wittgenstein, and Husserl, to name but a few) but they could not conceive such a vision when the foundations of physics — and particularly quantum physics— were too much uncertain yet or too radically new. One may hope something better from the 21st century, because one needs absolutely a new philosophical conception of nature in real accordance with science.

In keeping with this somewhat provocative opinion, which expresses basically a strong desire, I must say that many physicists are interested in philosophy, many find nourishment in the present history of physics, which is excellent, but very few find it in the present philosophy of physics. I thought I had to mention this situation in this interview, not as a criticism but as pointing out an important difficulty, which should be analyzed to bring a remedy.

3. What aspect of the current work in physics can benefit the most from collaboration with philosophy?

Most probably, a better understanding of quantum mechanics (and more generally of the laws of nature) would draw such a benefit, which would be mutual. I believe however that this kind of collaboration should be extended to wider circles, including cognition sciences and mathematics. Let me explain this opinion by quoting a few of my own cherished ideas:

One now knows enough about the human brain and its evolution to understand the reasons of its agreement with Kant's restrictive rules on representation. These rules disagree however drastically with quantum laws and our language –including its philosophical version– cannot describe or discuss a quantum world faithfully. One needs cognition scientists to assess what the mind is able to comprehend, either from intuition or through ordinary dialectics, and what it means to understand formally, through mathematics.

An example of such a field for collaboration is a philosophical understanding of the superposition principle, including Feynman's version where everything possible enters in the addition of quantum amplitudes. Physicists know very well how to use it. They have also shown (with the help of mathematics) that these foundations imply the classical laws to which our brain is fitted (which is nothing less than deriving many features of Kant's foundations of philosophy from the quantum principles, by the way). But the contribution of physicists, as such, stops there. How can our vision of the world embody quantum superposition of everything possible and how Leibniz would have expressed it? What about the transmutation of the characters of physical laws at the junction between the microscopic and macroscopic domains? Many of us understand these questions at a technical level, but not in philosophy.

The superposition principle is basically a mathematical statement and, along a complementary direction, I believe that a triangular collaboration between philosophers, physicists and mathematicians could be very fruitful. I explained in a recent book why times are ripe for a renewal of this convergence. From a fundamental standpoint, the main question is concerned with mathematics and physical laws, both in their respective nature and their relations. From an epistemic standpoint, one can wonder how the language of mathematics could enlarge the language of philosophy, as it does for physics, when it encompasses some concepts and laws that are so formal that no ordinary words can express them and no non-mathematical argument can fully envision their consequences.

4. What area in contemporary philosophy of physics is most fertile?

I do not know the contemporary philosophy of physics well enough to answer that question. The books from which I learned most were concerned with the philosophical aspects of cognition sciences, which is a very fertile area, and with the philosophy of mathematics, which remains fascinating.

5. Which area of physics holds the most exciting promise in the coming decades?

I suppose these promises are concerned with future work in philosophy. In more prosaic terms, I understand it as: "What topics are worth a special attention in philosophy courses?" The drawback is that most people answer this kind of question according to their own preconceptions. Anyway, do not take it as a joke when I say: More than enjoying exciting academic promises, everything dealing more or less with ethics and the future of Mankind, including survival, is a matter of duty in the coming decades. Many areas of physics are directly concerned with such an endeavor, but they differ from past ones. One may wonder for instance whether the legitimate fascination of 20th century for fundamental issues in physics will not fade slowly away, at least for some time, and whether philosophy will not come back in its own right to the first place.

6. How was I initially drawn to the field and what are some examples of my work that influenced the discipline?

As a student (say under 20 years), I was equally attracted by mathematics and philosophy. Then I discovered physics and I found it even more fascinating. I used much mathematics in my work as a physicist, but philosophy remained more than a hobby during most of my life. I knew Althusser and some other philosophers, but I was more attracted by the "spontaneous philosophy of scientists" than by postmodernism. In other words, I remained an interested layman in philosophy.

Late in my career, I was lucky enough to participate in a renewal of theoretical research on the foundations of quantum mechanics. I worked on its three main aspects, namely the derivation of classical physics and determinism, logic in the language of consistent histories, and the theory of decoherence, so that I was led to write some review articles and then some books, on these topics and their consequences. Since old and legitimate philosophical questions surrounded this domain, I was led to express some ideas about the meaning of recent advances and I became then almost completely absorbed by the depth and beauty of their philosophical aspects. I cannot appreciate however whether my work had an influence on the discipline or whether it had none, but who minds? My motto is still Husserl's, or characteristic of an amateur: Only the thing itself is worthy of interest.

14

Carlo Rovelli

Centre de Physique Theorique de Luminy
Université de la Méditerranée, France

---

1. What is the relationship between philosophy and physics? What should the relationship be?

The relation between physics and philosophy is much stronger than most physicists and most learned people generally assume today. In the past, this relation has been very effective and explicit. During the second half of the 20th century, the relation has been largely denied, for reasons that have to do both with the internal evolution of physics and the internal evolution of philosophy –reasons that I try to describe below, and that I think are no longer valid.

2. How did philosophers contribute or fail to contribute to the development of physics in the 20th century?

The evolution of fundamental physics has been very strongly influenced by ideas and arguments deriving from philosophy, during all the major advancements of the discipline, in all centuries. Galileo, Newton, Maxwell, Botlzmann, Heisenberg, Einstein, just to name a few major names, are explicit in their writings in recognizing the influence of philosophical ideas on their scientific achievements.

In the 20th century, for instance, Heisenberg would have never introduced some of the basic ideas of quantum theory, ideas that we use heavily today, if it wasn't for the anti-realist philosophical climate in which he was immersed. Einstein writes repeatedly that he has been able to overcome the Newtonian conception of space largely thanks to the critical tradition that goes through Leibniz, Berkeley and Mach and which has an exquisitely philosophical character. Recently even a strong influence of a philosopher that might appear above any suspicion of having to with physics, Schopenhauer, has been nicely evidenced in Einstein's conceptual struggle to build General Relativity. And so on, the list of examples might continue forever.

These direct influences are much stronger in the "revolution" periods during which fundamental science modifies and re-adapts its own conceptual structure, as the examples illustrate. In the periods of more "normal" science, where some basic grammar for describing Nature is established, and the struggle is concentrated into applying this grammar for describing more and more phenomena, physics is much less in need of philosophy. This is what has happened during the second half of the 20th century: quantum mechanics and relativity have been such a major conceptual steps ahead that the urgency was to apply them, to exploit their extraordinary novel power. This has lead to atomic, nuclear, condensed matter, particle physics, to cosmology, relativistic astrophysics, etcetera. In developing this wide spectrum of application of the new ideas, physics has taken increasing distance from the conceptual struggle that characterized its early 20th century major advancement, and has moved away from philosophy. To this has also contributed the move of the main centers to the USA, with its more pragmatic philosophical mood.

3. What aspect of current work in physics can benefit the most from collaboration with philosophy?

Today, the push to apply quantum theory and relativity to larger and larger domains is decreasing, and the problem of combining quantum theory and general relativity has taken the center of the stage. This means that the conceptual issues are once again on the table. Physicists are once again facing basic conceptual problems such as What is Space?, What is Time?, What is an observer?, Can we model the entire reality?, and so on. The conceptual structures on which quantum theory and general relativity were separately built does not suffice anymore for understanding the two together. Therefore physicists are forced back to the kind of investigations where philosophical inputs are essential. To the style of investigation which was the one of Galileo, Newton, Faraday, Heisenberg and Einstein: asking foundational and conceptual questions; not just applying a powerful conceptual structure, but modifying this conceptual structure. This is why an increasing number of theoretical physicists start again taking philosophy seriously, reading philosophy and talking to philosophers.

4. What area in contemporary philosophy of physics is most fertile?

The other side of this story of the 20th century divorce between science and philosophy is the internal developments of philosophy itself. It takes two to divorce. 19th century positivism fell on the trap of misunderstanding the immense success of Newtonianism. The faith in a scientific "Truth" found once and for all, has nurtured the idea of the effectiveness of a purely scientific path to knowledge, and, related to this, the dismissal of any "philosophical" enquiry. The legacy of the mistrust towards "philosophy", or the mistrust towards "metaphysics" has been immense. It has reverberated across the logical positivism and is still very widespread today. Many physicist's still get nervous at any appearance of the word "ontology"; and only last week I received a mail from a friend, a philosopher, who is organizing a conference on the interpretation of quantum mechanics, and who writes that he reserves to himself the right of ringing a warning bell during the meeting at any slide into "metaphysics" –in the pure Vienna-Circle tradition of shouting out "M!" at any hint of the "despised Metaphysics". But with all the large amount of garbage that the Vienna Circle (beautifully!) wanted to get rid off, perhaps something useful has fallen away as well, in my opinion.

Oversimplifying, philosophy has then bifurcated into two major rivers, whose influence is extremely strong in modern society. On the one hand, some continental philosophy has rejected the anti-metaphysical stance on the basis that what was thrown away was far more valuable for the human beings than the "arid" techno-scientific knowledge; this is nurturing much of the silly contemporary anti-scientism, with influences that reach the American legislation. On the other hand, large sectors of analytic philosophy have taken an over-respectful and subsidiary position with respect to science, reserving to themselves an ancillary role of description, comment and clarification, but leaving science alone to search for knowledge.

The fall of Newtonianism at the beginning of the XX century has came as an immense shock, and, to some extent, much of the later philosophy of science is a struggle to make sense of a truth –scientific truth– that can be as effective and "correct", and at the same time as "wrong", as Newton's theory. And of a form of enquiry –scientific enquiry– that can be find conceptual grounds as solid as Newton theory, and then through them away for new grounds. It is increasingly clear that the scientific enquiry is a complex one that cannot be easily codified. An essential part of it is formed by the construction of conceptual structures in terms of which to frame and organize the information we have about the world. These conceptual structures are not stable. They evolve with time. Science is a complex evolving game in which the rules of the game themselves evolve with time. Strict determinism appeared to be one of the absolute rules of the game (what is scientific knowledge if the same causes do not produce the same effects?); still, it is no more part of fundamental science, after quantum theory. Writing equations describing how the state of a system evolve in time appeared to be one of the absolute rules of the game; still this is not the way the world is described in general relativity. Observer-independence of the phenomena appeared to be one of the absolute rules of the game; but are we truly sure quantum theory is compatible with it? And so on. These burning issues, in my opinion is where science needs philosophers to think.

In trying to finally come to terms with the full scope of the discoveries of the 20th century, fundamental physics is raising issues that are acutely philosophical. Physicists need to create new conceptual frames in order to make sense of the phenomena. But creating, testing and comparing conceptual frames is what philosophers have been doing for centuries. So, physicists ask for help from philosophers. Nothing strange in this: they start again doing what Newton, Einstein and Heisenberg used to do. Philosophers, on their side, are today a bit surprised. They react as saying: "you are the ones who are supposed to find out the world; my role is only to comment". I think they are missing something. Some of the burning questions at the core of quantum theory, quantum gravity and general relativity need the special clarity of philosophical thinking. Physicists expect from philosophers new ideas, substantial criticisms of imprecise assumptions, general points of view of the world. That is, the kind of nurture that physics has found in the thinking of the philosophers all across the centuries.

5. In your opinion, which area of physics holds the most exciting promise in the coming decades?

20th century physics has opened a great scientific revolution, where our understanding of Nature has been modified in depth, perhaps as much as during the Copernican revolution. But the revolution is not finished yet, because we have not yet reached a synthesis where everything we have learned in the 20th century makes consistently sense. We are half way through the woods. The Copernican revolution took 150 years to reach its end, in the final synthesis of Newton's Principia. The present revolution was opened at the beginning of the 20th century. In my opinion, the most exciting promise in the coming decades is the completion of the revolution. A new synthesis in which the fundamental discoveries about the basic grammar of the physical world made in the 20th century, quantum theory and general relativity, will make sense together.

Careful, this has nothing to do with the "Theory of Everything" or the "Unification of all the forces". The dream of the "Theory of Everything", the universal formula that describes everything, has always been there in physics, has always appeared just round next corner, and has never been realized. I think we are very far from it. I think there is a good chance the search for it might fail. What I do expect, on the other hand, is that in the coming decades the 20th century revolution could be completed; that is, the conceptual consistency that has characterized physics since Newton, and was lost in the 20th century, would be regained.

6. How were you initially drawn to the field and what are some examples of your work that have influenced the discipline?

What has soon fascinated me in physics is not problem-solving or the pleasure of intellectual games. It is the fascination of a way of thinking that is capable of modifying itself, or re-drawing the basic map of the World in front of our eyes. I see the intellectual adventure which is the search for knowledge, the rational investigation of reality, as one of the most fascinating enterprises of humanity. I took as a privilege the possibility of trying to give my own tiny contribution to the common trip.

Loop quantum gravity, which I have begun to develop with Ashtekar and Smolin in the late 80's, has given a new impetus to the struggle of merging the conceptual novelty of general relativity with quantum field theory. In my work, I have always tried hard not too overlook the basic conceptual questions; I think that it is in reflecting on these, not on technicalities, that fundamental science has mostly advanced in the past, and will in the future. My book on quantum gravity is written very much in this spirit, and I have tried to influence my discipline in this direction.

15

Lawrence Sklar

Department of Philosophy

University of Michigan, USA

I

Foundational physics is a curious discipline indeed. Physics rests, of course, upon the accumulation of observational data, with ever more refined methods for empirically probing into nature requiring ever more elaborate theories to predict novel observations and to explain the existing known general structure of the observational results. And the project of finding such predictive and explanatory theories requires that striking and mysterious ability of the part of theoreticians to imagine ever more elaborate and ingenious theoretical structures. Indeed, the most intractable part of any sociology or psychology of science is that which attempts to give us at least some partial understanding of how such novel explanatory structures can arise out of the scientific imagination. Once a theory has been posited to deal with some realm of experimental data, there is, of course, the question of whether or not there is sufficient warrant for the theory in the evidence to lead us to "accept" the theory into our corpus of currently established theoretical scientific beliefs.

Now no matter what the science is, this simple model has been subjected to many philosophical arguments designed to convince us that we must be cautious in our epistemic enthusiasm. Problems of evidence abound: issues of error, of delusion, of the limitations on our abilities to access the world, of implicit bias and partisan selection in our evidential searches, are familiar issues in skeptical philosophy. And when it comes to going beyond evidence to theory we have all the familiar skepticism framed around the issues of the justification of inductive or abductive methods, again of implicit bias and pre-selection, of the problem of all the as yet unimagined hypotheses we have ignored, and so on.

On top of these grounds for skeptical doubts about the "believability" of our best scientific theories there is also the famous induction from historical experience that tells us that we have good reason to believe that in the future we will no longer accept the theories we currently take as our best available science. How can we, then, now claim a good reason to believe in these theories. Even more, there are the arguments to the effect that even our best theories can deal only with highly idealized systems, and therefore cannot be taken as truly characterizing the real systems of the world.

But these issues, common to all of the sciences, only tell part of the story when foundational physics is concerned. For it is in that area of science where we run into an additional problem: We have no definitive idea of what our best accepted theories say at all. It isn't merely that we need some refined doxastic and epistemic terms to describe the attitudes of belief we ought to take toward these best foundational physical theories. We need some way of deciding just precisely what these theories claim to be telling us about the nature of the world. This is the problem of interpretation.

II

Every foundational theory in the history of physics has thrown up a raft of interpretive questions. These have led to seemingly interminable debates about just what the theory is saying about the fundamental ontology of the world.

In classical physics, are forces nothing but fictitious features representing intervening variables? Or, rather, as Boscovich contended, are forces the fundamental constituents of the world? Does the theory tell us that space itself exists as an entity over an above the material constituents of the world and their relations, or is it a theory just about such material things and their relations? Does it, or does it not, tell us that some absolute measure of the lapse of time exists? Can a relationist account be maintained if we adopt some cosmological version of a Machian theory, and, if we do, is the cost of that (the posited no absolute rotation for the cosmic matter) too high a price to pay? And what are we to make of the least action formulations of the theory with their seeming appeal to teleology and to events in the present being accounted for by reference to both the past and the future of the system in question?

Add a theory of universal gravitation and all the questions about the legitimacy of action-at-a-distance theories come up. Can gravity be properly understood only if there is some hidden, "mechanistic" account over and above the Newtonian mathematical formulation?

Move to the special theory of relativity and we encounter the never ending debates about the "conventionality" of distant simultaneity (and what on earth that might mean). The general issues of conventionality also arise with the questions about in what way, if any, special relativity is superior to oner of the available (properly understood) aether theories. Go on to general relativity and more interpretive questions pile up. What happens to the old substantival vs. relational debate about spacetime in the presence of the dynamically changeable and interactive spacetime of general relativity? How does Einstein's notorious "hole" argument affect that debate?

In the realm of thermodynamics and statistical mechanics we have all of the deep and persistent questions about the origin and justification of the probabilistic posits needed to ground the theory. Do they represent some basic, independent feature of the world? Or do they have a cosmological account as being just one sample of an infinite realm of cosmic universes? Or are these probabilities subjective in nature? Where does the time asymmetry of the thermal world come from? From some as yet not fully understood time asymmetry of the fundamental laws (from GRW quantum mechanics of the basic constituents of systems, for example), or from some initial absurdly low probability initial state of the cosmos at the Big Bang? Is thermodynamics truly reducible to some underlying theory, or are its basic principles assumed in trying even to frame such a theory?

Go on to quantum mechanics and, of course, all hell breaks loose! Now interpretations become rampant and disagreement astounding. Is measurement just a reflection of the many degrees of freedom of the measuring apparatus, or does it show a true hiatus in the usual deterministic evolution given by the dynamical laws? Should we fit the quantum mechanical predictive probabilities into an underlying deterministic theory of particles and ghost fields – even at the price of instantaneous action-at-a-distance and the denial of the equivalence of inertial frames at a fundamental level? Or would we be better off positing a universe that is constantly branching into a multiplicity of universes with every possible outcome of every possible measurement realized? Or, heaven help us, do we need to posit transcendent egos acting upon the world as well as being acted upon by it to do justice to the projection postulate of the measurement process?

Advance to quantum field theory and all of the interpretive issues of ordinary quantum mechanics remain, but now have superimposed upon them a raft of new problems. How are we to understand the existence of multiple, non-unitarily equivalent representations compatible with the commutation relations in this realm of entities with infinite numbers of degrees of freedom? Can we think of the theory as being one of the fundamental particles of the world, when it seems to show that nothing localizable in the manner of ordinary particles can exist and when it posits a relativity of the number of particles in the world to the state of motion of the observer? Will an ontology of field fare any better? If not, can we find our way out by some localized version of the theory such as local algebraic QFT framed from the beginning to avoid some of the mine fields of the traditional theory? If we find the theory afflicted with non-physical divergences, and we scheme to avoid those by such peculiar gimmicks as renormalization theory, does that tell us that a new way of applying theory to the world is at hand, or only that we are temporarily forced to resort to methodological tricks that have no place in fundamental theory in the long run.

Then try to combine the insights of quantum mechanics and QFT with those of general relativity, only to discover new problems about how to even begin to understand the place of time as a fundamental parameter of our theory.

III

The philosophy of physics will continue to struggle with all of these interpretive issues. One overall aspect of the issue of interpretation of foundational physical theories seems clear. It is hard to think of any case where we can say that an interpretive problem has been "solved." Rarely if ever do we find questions passing into oblivion in the face of universal (or near universal) agreement that a final resolution of the issue has been obtained. Rather it seems as though the old questions remain open and lively as new ones arising out of new physics join them in the realm of interpretive puzzles.

What does happen is that, in the light both of new physics and in the light of interpretive progress, the old puzzles continually metamorphize into new versions of greater sophistication and greater subtlety. The now ancient issue of substantivalism vs. relationism in spacetime theories has not disappeared. Instead, both in the light of the sophistication of physics in the form of general relativity, and in the sophistication of philosophy, say in the newer understandings of modal discourse or in the newer versions of Machian approaches to relationism, the old issues remain open in novel dress.

So philosophy of physics will continue its fruitful and exciting course of exploring each of these interpretive issues, old and new.

But there is something else to be done that has not been explored to anywhere the degree it deserves. This is the general problem of understanding interpretation itself. The very idea that there is something that foundational physics demands beyond the search for empirical data, beyond the imaginative formulation of hypotheses, and beyond the program of exploring the degree to which hypotheses stand up against empirical test, is puzzling and problematic. Why is there a need for something we think of as interpretation? Are there general issues of interpretation that cut across the specific theories and specific interpretations that are the daily bread of philosophy of physics? Are there general methods of carrying out interpretations that, again, appear over and over again in a wide variety of interpretive explorations? Finally, what should our doxastic and epistemic stance toward foundational theories be, given that they seem to abide in a permanent state in which multiple, apparently incompatible, interpretations remain in contention?

What opens up the demand for interpretation in the first place? In general it is some feeling that something about the foundational theory is problematic. This may be as narrow as a puzzle about the mathematical legitimacy of some component of the theory, as in problems with Heaviside's methods or delta functions in quantum mechanics, or, more deeply, worries about the renormalization techniques of QFT. More interestingly the puzzles may be about some "metaphysical" aspect of the theory that seems to some unsatisfactory. Consider, for example, the relationist's worries about Newtonian substantival spacetime or the Cartesian mechanist's rejection of Newtonian gravity as action-at-a-distance. Similarly think of the interpretive puzzles raised by entanglement in quantum mechanics where our ordinary notions of how causal influences work are deeply shocked by the dependence at a distance of components of entangled systems.

In other cases there are features of the theory that seem to suggest that other features of the theory are failing in some sort of legitimacy. In statistical mechanics, for example, the underlying dynamics generally is presupposed to allow for any kind of initial conditions whereas the theory as a whole posits a quite mysterious probabilistic constraint on how the initial conditions of a set of systems of a given macroscopic kind must be distributed. In an even deeper problematic case, the purported universality of the dynamical rules of quantum mechanics seem to leave no room for the necessary non-unitary measurement process, or even, indeed, to allow us to characterize in clear cut physical terms just what processes in the world are to be counted as measurements.

Interpretation begins with a need, a need to come to grips with some aspect of a foundational theory that feels "wrong" or "defective" to us. A question well worth exploring is whether there are general categories of such "defects" that encompass specific problems from a wide variety of foundational theories.

Once a problem has been encountered, attempts to deal with it are the meat of interpretation. What moves can we make to draw the sting of the conceptual difficulty? Here again we encounter a fascinating mixture of the specific and the general. Each foundational theory with its specific kind of conceptual problem requires its own particular set of interpretive programs. But we can discern some general patterns in interpretation that run across a variety of foundational theories.

Interpretive moves have a very wide range. At one end there are the elegant ways in which improved mathematics can alleviate the problems of non-rigorous mathematics in the theory (for example, use of distributions or spectral decomposition to get rid of delta functions). At the other extreme there are those proposals that would replace an existing theory with its conceptual problems with what looks like a novel theory – a novel theory allegedly solving the problems. Here the novelty of the new theory is indicated by the fact that it even makes distinct empirical predictions than those made by the problematic theory. The GRW interpretation of quantum mechanics would be an example of this.

In between are all those interesting cases where interpretations are given that allegedly help with the conceptual difficulties, that require more than some refined mathematics, but that leave the empirical predictions of the theory in question intact and are best thought of as interpretations of the theory rather than replacements to it. In the case of quantum mechanics, for example, this would include such broad ranges of interpretation as Bohr's

Copenhagenism, many-worlds interpretations, Bohm's ghost field hidden variable theory and modal interpretations.

But, once again, a question well worth exploring is whether or not there are general modes if interpretation that appear again and again across a wide variety of interpretations from a wide variety of foundational conceptual theories. One such mode that comes to mind is that which tries to evade the conceptual difficulties by some kind of empiricism or positivism. Focus on the empirical predictions of the theory, try to locate the conceptual difficulties at a non-empirical level. Then try to interpret the theory in an empiricist or positivist way as a device for establishing the correct correlations among possible measurements or observations, thinning out the theoretical ontology to avoid encountering the conceptual problems. Within this category of interpretations are "moves to the local," which find the conceptual problems to have their origin in various global or non-local aspects of the theory in question, and seek to resolve the problems by eschewing such too rich, non-local, concepts. Interpretations as disparate and facing radically different conceptual issues as relationism in spacetime theories and local algebraic interpretations of quantum field theory can be illuminated by looking at these interpretations from this perspective.

And so we have the general problem: What kinds or sorts of general approaches to interpretation can be found that cut across the specific interpretations idiosyncratic to particular foundational theories with their specific conceptual problems?

IV

Finally, we need to think a lot more about just what the implications of the existence of multiple interpretations are for our doxastic and epistemic attitudes toward our best foundational theories.

We know of a number of reasons that have been espoused for dropping any simple-minded idea that we ought to "believe" our best theories, or take them as "true" of the world. Our theories are transient and will sometime be replaced in our domain of accepted accounts by novel accounts not strictly compatible with our current theories. Our theories deal only with idealized representatives of real systems in the world, and not with actual systems in all their messy reality. The confirmational relation of evidence to theory is at best probabilistic. Other, less promising, grounds for varieties of skepticism have been proposed as well.

But multiple interpretation presents it own challenges to simple belief models. How can we adhere to the idea that we ought to believe what our best available theories in foundational physics tell us about the world, when we don't ever have a genuine consensus with regard to just what these theories are claiming the world to be like?

The usual alternatives to a simple belief model spring to mind. Maybe we should go positivistic and take the theories to be reducible to congeries of their empirical claims. Or perhaps we should be "constructive empiricists" and take the theories to be realistically construed but not care about just which interpretation of them to believe. Or perhaps we ought to believe that in the long run the dilemmas posed by multiple interpretations will cease to haunt us, since in the fullness of time science will fix on some one best interpretation for each part of its best available foundational physics, just as in the progress of science we will fix on some permanent best accepted physical theory – a version of Peirce's "faith and hope."

But these usual responses leave us with a sense of disappointment. For one thing not all of the multiple interpretive possibilities can be cashed out in terms of some unique equivalence class of nicely characterized "common empirical predictions." And we would like to have some appropriate doxastic or epistemic way of characterizing what attitude we justly have to our theories and their interpretations now, and not just in some perhaps never to be obtained future when all dispute has handily vanished away.

But for now what I want to emphasize is that along with the continued deeply profitable exploration of the specific possible interpretations we can discover for the individual foundational physical theories we have and for their idiosyncratic conceptual difficulties, a general philosophical exploration of why interpretations are needed, how they are generated and assessed, and how we are understand our attitude toward our present best foundational theories and their multiple interpretations is in order.

16

Paul Teller

Department of Philosophy

University of California, Davis, USA

I was posed six interview questions. I did not like some of them, especially ones asking about "the most" of something, such as "which area of physics holds the most exciting promise in the coming decades?" Just why I resist such questions will be part of my story. Since this is an "interview", I will begin with a sketch of how I came to one area of work that I perceive to be of exceptional importance to our understanding of science and, indeed, the whole philosophical enterprise.

As an undergraduate I thought I wanted to become a scientist but quickly realized that signing on to any one area would condemn me to the life of a narrow specialist. One day it hit me, like receiving my vocation, that by becoming a philosopher of science I could focus on the questions that really interested me about the form, the content, the sweep of our larger understanding of the world. As it would turn out, my mature work would increasingly sharpen its focus on the nature of that understanding itself.

For my thesis and in the immediately following years I worked on confirmation theory, but soon came to feel that neither I nor others had worthwhile new approaches to its knotty problems. So I turned my attention to the interpreter's "big one": quantum theory. Here I want especially to mention how in 1976 I worked through all three volumes of the Feynman lectures. I became increasingly puzzled. Throughout, Feynman developed the art of constructing fabulous stylized simplifications to do duty for the complications of the messy real world. Where were the Positivist's deductions from natural laws? I filed all this in my tacit anomaly drawer and joined the wresting match with the puzzles of quantum theory. I knew enough physics that I should have known better. But that is hindsight. In the 1970s we were all stuck.

In 1983 I took my next step away from the picture of physics as a deductive framework, but again one that I can only see with hind sight: my review for Nancy Cartwright for her promotion to full professor. The page proofs for her How the Laws of Physics Lie provided the centerpiece of her file, pages that provided difficult reading. I absolutely did not get it. But I did get it that I didn't get it, and I got it that it was extremely important. I wrote that in this book Cartwright had "left the profession with a clearly delineated challenge", that of coming to terms with her case that the laws of physics lie; and I concluded that "I can think of no problem which is either more difficult or more important for understanding the nature of science." I supported her promotion in the strongest terms.

I don't know how long it would have taken me to "get it" were it not for conversations with Ron Giere. I remember lunch with him in Greektown just nextdoor to my department at Chicago Circle. He got all excited as he regaled me with the ideas he was putting together in writing his Explaining Science.

Slowly I started to see the point that with today's familiarity we can state with greater ease. Physics, and science generally, isn't in the business of deduction from exact true laws. It is in the business of crafting models that are always limited in scope and, even where they apply, always limited in their accuracy. The familiar laws function as model building guides and, as Ron kept insisting, are exactly true only in the idealized models that they help to define. The models then do their representational work when we make use of their agreement with the world, always in limited respects and with imperfect accuracy. This is what today I like to call "the model view" or "the modeling attitude". It was not until much later that I began to see the significant overlap with Cartwright's work.

I think I got it, but I think I only gradually got how important it was. I was struggling to get my head around enough of quantum field theory to begin working on its interpretive problems. During this period my appreciation of the predominantly modeling aspect of theories grew steadily, and it played a muted but important role in my 1995 An Interpretive Introduction to Quantum Field Theory. In the succeeding decade the modeling view and its repercussions have attracted ever more of my attention. To explain why let me turn from my own to the larger story of the vicissitudes of what I like to call "the perfect model model".

Throughout the 18th and 19th century many interpreted Newton's legacy as the discovery, by him and his successors, of exact laws that were supposed to govern the operation of the cosmic machine. A corralorary took physics as the "fundamental science", the science that would tell us, in principle if not in practice, how everything works: The copuscularian philosophy of the scientific revolution had taken all natural phenomena to be comprised by the shaped matter in motion of the most minute constituents of all bodies. Consequently, determination of the laws that governed matter and its motion would, as Laplace put it, enable a super-intelligence who was cognizant of the world's present mechanical state also to see its entire past and future. In the late 19th century physicists such as Thomson and Michelson apparently saw physics as essentially done - only the details needed filling in. This was to be the perfect model of the world. The picture was itself, of course, a model of the scientific enterprise. Hence the epithet, "the perfect model model"

Not all adhered to the perfect model model. There was not only the 18th century romantic reaction to Newtonianism; 19th century physicists such as Boltzmann and Hertz expressed views that included striking common ground with contemporary modelers. But somehow - and there is an interesting story to be told here that I do not know - by the second quarter of the 20th century the ambition of fundamental physics as the comprehensive and exact theory of the world appears to have eclipsed modeling attitudes. The refinement of the work of Newton and his successors with quantum mechanics and the relativity theories should have bred a more cautions attitude; but instead a great many physicists and interpreters of physics thought that the perfect model was at last almost within our grasp. This attitude was widely echoed more broadly within science and extensively in science's larger popular image.

By the last quarter of the 20th century the situation had become more complicated. In many quarters fierce reductionist attitudes predominated, for example in the arguments by molecular biologists that other branches of biology would become obsolete. Fundamental physicists, while recognizing the practical impossibility of comprehensive reduction, clamored for funding for the super-cooled super collider that would take us to the "final theory". But physicists in other subdisciplines - especially many body physics - began to give explicit voice to what they knew from their own work, that physics is in the inexact model building business.

This was the point - 1983 - at which Cartwright's book began a tide of explicit critical examination of the perfect model model, a tide that has swelled with the further work of Cartwright, Giere, their students, and many others. Modeling figures everywhere in physics, not just in subjects such as the study of materials, acoustics, and geophysics, but prominently in quantum field theory and general relativity. NOWHERE in physics do we have laws that are both completely precise and completely accurate. For example, the speed of light is a constant, c - in a perfect vacuum. Conservation laws are tied to symmetries - that are nowhere perfectly instantiated in our messy real world. What goes in physics goes throughout science. Scientific accounts work ubiquitously through inexact models with limited application. There are many unqualified truths, but none that are also completely precise. While I would not want to insist that these claims hold with no exceptions - the account is, after all, itself a model - exceptions are hard to come by.

I don't think that any of the foregoing is news to scientists who have all along understood this aspect of their craft, always implicitly and usually explicitly. Consequently we should not expect these realizations in the sphere of philosophy to have substantive repercussions for the practice of science. But appreciation of the modeling nature of science radically revises how we understand science and how we go about analyzing it. For example confirmation theory can no longer be understood as the estimation of the probability of truth of theories - theories that we antecedently know to be false of the real world. We must look for ways of understanding the nature of scientific explanation without requiring the literal truth of explanans. Not as obvious, but shifts that I believe we will see, are ones in what is at stake in the realism debate and the relation between -sometimes conflicting - theories. There are many shifts also in more specific issues. Take, for example, the discussion in the 1970s and 80s, continuing today, about whether there are any laws in biology. When we appreciate that all laws in physics are true only of idealized situations and apply to real cases only with qualifications, the claimed contrast with generalizations in biology needs to be rethought.

We are just beginning to see the fruits of the modeler's reorientation towards problems in philosophy of physics and in philosophy of science more generally. I suspect that these redirections in philosophy of science are just the beginning. When one remembers that science is supposed to provide the sterling best of human knowledge and representation and then take to heart the modelers limitations, how much more must these limitations apply to human knowledge and representation more broadly? The basic guidelines of the modeling approach apply to any subject matter (with possible exceptions in mathematics). For any given subject matter we should not expect to find exact truths - this is rejected as not feasible in practice. Instead we seek inexact models that characterize the subject matter at hand, just as we do in science. When one accepts this attitude as a general guide, we must recharacterize how we evaluate our conclusions. We evaluate the fruits of our intellectual labors not for exact truth but rather for success in achieving specific practical and intellectual objectives. The relativization of evaluation to objectives means that the evaluations themselves are likewise relativized. A model that succeeds in one objective may fail in another, while another model may succeed where the first failed and fail where the first succeeded. As a result pluralism becomes a real option. Any pluralism is heavily constrained. There is - always - the contrast between success and failure in achieving our objectives. But with a plurality of objectives comes a concomitant relativization of evaluation.

All of the foregoing considerations have repercussions for the practice of philosophy. Starting with public philosophical debate, we still see a great deal of the standard to which I was trained: "Here are 17 flaws in your account. So your account is false. Oh yes, you point out 23 flaws in my (competing) account. I'm working on them." This is crazy! Instead every account should be critically examined for both its strength and weaknesses and compared in light of these. Many accounts will have so many defects in comparison with its virtues that we will sensibly abandon it. In some cases one account will come out ahead of a competitor on any measure. But in many cases one account will be judged good for this, not so good for that, while a competitor will have complementary strengths and weaknesses so that both will be found worthy of further study, correction, and refinement.

The modeling attitude encourages its own approach to doing philosophy. We renounce the hubris of discovering God's laws, the exact structure of reality. Instead we work as artisans, individually and collectively, fashioning intellectual tools and artifacts that facilitate human understanding of those parts and aspects of the world that we can partially and imperfectly discern. Each tool will have its strengths and weaknesses. We need always to keep our objectives in mind: the phenomena of current interest, accessibility of the ideas we are fashioning, ironing out puzzles, forging links with other phenomena, and so on. We struggle to balance success in such objectives against distortion and other defects. So doing always involves trade-offs - and how we chose those trade-offs always involves elements of personal taste.

Such an approach to doing philosophy applies not just to our work individually, but to doing philosophy with others. Often work with others follows the old pattern of public philosophical debate. I point out to you difficulties with your account, and you point out to me difficulties with mine. Usually this is very useful - often we need another pair of intellectual eyes to see problems to which we are blind. At the same time such philosophical sparing contrasts with the exceptional thrill of "doing philosophy". You and I become jointly engaged in "figuring something out". We are both critics and both authors of new ideas and arguments. We help one another in keeping track of both the objectives and the strengths and weaknesses of our creation as it takes shape. In particular we are guided away from seeing every limitation as a fatal flaw, the quicksand of traditional philosophical debate. In such ways we enter into the collective process of fashioning an intellectual structure that is not mine nor yours but ours.

Usually such work requires that you and I share a good deal by way of our familiar analytic tools, principles of model construction, and standards of success. There are also cases in which the problems on which others work and the standards of success that they apply are so at odds with our own that the best for which one can hope is respectful disagreement. But sometimes there is much we can gain by "visiting a foreign philosophical land", much profit that we can find by trying out, for the day as it were, even radically different ways in which others work. We put on their thinking cap, we take on their problems as our own, and learn, as through their eyes, their methods and standards. We do philosophy with them in their way. Many times we learn a great deal from such a visit that remains of value to us when we return to our own familiar landscape. Sometimes, when we have returned home, we realize that the gulf that we had perceived was not quite as wide as it had seemed.

The modeling attitude was reborn in recent decades in the recognition that, in practice, even foundational physics, the "first science", does no better than trade in inexact, limited models. This new attitude is having profound repercussions for how we pursue problems in philosophy of physics, in philosophy of science more broadly, and, really, in all of philosophy and beyond. We can look forward to all of our intellectual pursuits embracing a critical pluralism, with imperfect representational tools to be evaluated in light of the work to which they are put.

And, oh yes! I said that my story would include an explanation of why I reject questions phrased in terms such as "the most" -" What area in contemporary philosophy of physics is most fertile?" - and the like. I hope that by now the explanation is obvious. The most for what? My theme has been that one achievement will be good for some things, another achievement good for others. I will certainly apply that to what I have suggested here. The achievements of modelers, already realized and to come, are among the recent important contributions of philosophy of science; contributions on which I have focused because they have so much captured my own imagination.

17

Steve Weinstein

Department of Philosophy
University of Waterloo, Canada

1. What is the relationship between philosophy and physics? What should the relationship be?

Many philosophers of physics seem to be interested in "interpreting" what they think of as "our best physical theories". They think that fundamental physics has ontological lessons for us, and they furthermore seem to think that these lessons have yet to be properly extracted. Some of my best friends and colleagues work along these lines some very smart people, but I think they're wrong. I think that, far more than we're willing to acknowledge, what you see is what you get. I think there's useful work to be done in understanding what's in front of us, but I don't regard this as interpretative work.

I think that philosophers might more usefully investigate concepts that physicists take for granted, but really do not understand. What does special relativity have to do with causation? There's a principle of "relativistic causality", but why does it bear that name? What is it to explain the arrow of time? What is the arrow of time? Some people working in these areas are, I think, doing really interesting work. It's not surprising, because there's so much to work with.

I think that physics has had real consequences for ontology, for metaphysics. The idea that simultaneity might be relative is truly astonishing, and it is an extraordinary contribution to the philosophy of time. What I'm skeptical about is that there's much in the way f interpretative work left to be done.

3. What aspect of current work in physics can benefit the most from collaboration with philosophy?

Thermodynamics and statistical mechanics, in particular attempts to "derive" thermodynamics from statistical mechanics. Jos Uffink, among others, has done excellent work in this area.

4. What area in contemporary philosophy of physics is most fertile?

I like some of the recent work on the role of explanation and approximation. And I think that modern cosmology, with its fine-tuning problems and occasional appeals to anthropic "explanation", is a good candidate for further philosophical investigation. I think that Landauer's principle and the problem of Maxwell's demon have seen some good philosophical work recently, and there's certainly further insight to be had in this area – I hope people get to it!

5. In your opinion, which area of physics holds the most exciting promise in the coming decades?

I think cosmology will be of continuing interest. There's a lot of data, the problems have great intrinsic interest, and they're attracting good people as a result, which should lead to more interesting observations. Ultimately, cosmology is the most likely source of new insight into fundamental physics of the quantum gravity sort. And I think it provides continued impetus to think creatively about the quantum "measurement problem", as it's the only physical context we have in which the measurement problem is a problem for physics itself.

6. How were you initially drawn to the field and what are some examples of your work that have influenced the discipline?

I was drawn to the field... I'm not sure how. I remember thinking about quantum mechanics when I was around 15, and coming up with a crude version of the many-worlds interpretation at the time. But I don't remember what I was exposed to that led me to think about quantum mechanics. I do recall, a few years later, enjoying Hilary Putnam's "A Philosopher Looks at Quantum Mechanics".

As with Bertrand Russell, Putnam's logic can be a little slippery, but the writing is enormously engaging.

As for my work, I don't think it has influenced the discipline! I have a nice little paper on Maxwell's Demon ("Objectivity, information, and thermodynamics") which might be worth the time of someone thinking about this topic. And I think my work on quantum gravity would be of use to anyone starting out in the field.

Index

Term Index